SIMULATING WORKPLACE SAFETY POLICY

Studies in Risk and Uncertainty

edited by W. Kip Viscusi
Department of Economics
Duke University
Durham, North Carolina 27706

Previously published books in the series:

Luken, R.: ENVIRONMENTAL REGULATION: TECHNOLOGY, AMBIENT AND BENEFITS-BASED APPROACHES

Shubik, M.: RISK, ORGANIZATIONS AND SOCIETY

Edwards, W.: UTILITY THEORIES: MEASUREMENTS AND APPLICATIONS

Martin, W.: ENVIRONMENTAL ECONOMICS AND THE MINING INDUSTRY

Kunreuther, H. and Easterling, D.: THE DILEMMA OF A SITING OF HIGH-LEVEL NUCLEAR WASTE REPOSITORY

SIMULATING WORKPLACE SAFETY POLICY

Thomas J. Kniesner
Indiana University

and

John D. Leeth
Bentley College

SPRINGER SCIENCE+BUSINESS MEDIA, LLC

ISBN 978-0-7923-9519-5 ISBN 978-94-011-0649-8 (eBook)
DOI 10.1007/978-94-011-0649-8

Library of Congress Cataloging-in-Publication Data

A C.I.P. Catalogue record for this book is available from the Library of Congress.

Printed on acid-free paper.

To Deborah Anne Freund
and
Lori Burr Leeth

CONTENTS

PREFACE

People want more from the government. One thing they want more of is a sense of personal safety, at home and at work (*Regulation*, Fall 1991). People also want the government to quit wasting money. The objective of having the government provide a safer life for us and our children at minimum cost leads logically to looking at policy within the system involving the private sector plus governments at the federal and subfederal levels. Using numerical simulations our book takes an integrated quantitative look at how the various institutions influencing workplace safety lead to the observed levels of illnesses and injuries among U.S. workers.

Our innovation is piecing together the mosaic of interactions among workers, employers, state government, and the federal government that is numerically realistic in the sense of using economists' current knowledge of quantitative connections. Our objective has been to write a Gray's *Anatomy*, if you will, of how the U.S. economic system, as tempered by government policy, jointly determines employment patterns, wages, and workplace safety levels.

The book that follows is best appreciated by a reader who likes numbers and is comfortable with algebra and multivariate calculus as they apply to situations where decisionmakers are optimizing subject to limitations. Because we try to show how numerical simulation can complement econometric studies a basic understanding of regression is also useful background for the reader. We have structured our presentations to be accessible and interesting to undergraduate honors, masters, and doctoral students in the social sciences, public policy, law, and public health programs. Although the data we apply are for the United States the underlying economic framework is valid for any economy so that the intellectual approach we develop should be of widespread interest. Our primary audience is academics and government policymakers trained in the quantitative social sciences who study and influence job safety policy.

The organizing scheme of our book is numerical simulation of how the labor market rewards workers in light of the fact people may be killed or injured at work. The book is chock full of numbers. The first set of numbers describes how a prototypical representation of the U.S. labor market would match workers to jobs in the absence of any government. We then added government regulation of the workplace characteristics and mandatory no fault injury insurance sequentially so as to see quantitatively both the separate and interaction effects of OSHA and workers' compensation insurance. The research objective is to learn the numerical effects of policy on wages, safety, and worker well-being. We emphasize that numerical simulation can fill in the gaps around econometric research to clarify why policy may or may not now be cost efficacious and to locate reforms that could be economically beneficial to workers' safety.

Our sequencing of chapters emphasizes the value of numerical simulation as a research tool from which policy relevant conclusions emerge. Chapter 1 highlights the data, public policies, and research approaches to studying workplace safety. Chapter 2 surveys the current econometric literature on the safety impacts of the two main government policies in the United States, federal OSHA regulations of workplace characteristics and states' workers' compensation insurance laws, identifying holes in our knowledge that could be filled in by numerical simulation. Chapter 3 sets up the labor market model for how numerical simulations can further our understanding of how policy influences labor market outcomes concerning workplace safety. After developing a model of the complete interactions of state and federal policies, we identify in Chapter 4 the degree to which policy coordination at various levels of government may be socially beneficial. Having earlier established quantitatively the dominance of workers' compensation insurance in the policy package, in Chapter 5 we take a deeper quantitative look at the connection between benefits, claims, and actual injuries to distinguish between a reporting and actual safety effect that has puzzled empirical researchers. Chapter 6 ends the book by taking what we have learned via numerical simulations and inferring a set of policy recommendations, including a controversial suggestion that the Occupational Safety and Health Administration be eliminated and the federal government concentrate on research and information dissemination concerning industrial hazards. We conclude with the recommendation in Chapter 6 that the states strengthen the economic incentives for safety via the legal system and the pricing of workers' compensation insurance.

We want our book's ideas to mindmeld with readers interested in simulating behavioral systems. We also hope to reach two other interested groups. First, we would like policymakers to see how numerical simulation of a complete set of policy interactions — main effects as well as side effects — leads to better use of scare government resources. Second, we hope to convince empirical researchers to add numerical simulation to their tool kits to supplement their econometric research methods.

In trying to convince the reader of the value of numerical simulation we offer concrete policies for more cost efficacious policies toward workplace safety. Our interest in numerical simulation began from trying to understand why OSHA has not had much of an impact on U.S. worker safety. The basic ideas in every chapter have been presented at professional meetings and published in refereed economics journals. Our arguments reflect a belief that while we acknowledge the value of qualitative theoretical research and econometric models they cannot always take the researcher as far into a topic as is needed. Theoretical results can be qualitatively ambiguous and data not rich enough to identify the complete economic structure for informed policy. We hope to convince readers that numerical simulation is a necessary tool to policy relevant social science research.

ACKNOWLEDGMENTS

We are grateful to the *International Economic Review* for permission to use material from Thomas J. Kniesner and John D. Leeth, "Simulating Hedonic Labor Market Equilibrium Models: Computational Issues and Policy Applications," *International Economic Review* 29(4), November 1988, pp. 755–790; © 1988; Economics Department of the University of Pennsylvania and the Osaka University Institute of Social and Economic Research; grateful to Cornell University for permission to use material from Thomas J. Kniesner and John D. Leeth, "Separating the Reporting Effects from the Injury Rate Effects of Workers' Compensation Insurance: A Hedonic Simulation," *Industrial and Labor Relations Review*, January 1989, 42(2), pp. 280-93, © 1989, Cornell University; grateful to JAI Press Inc. for permission to use material from Thomas J. Kniesner and John D. Leeth, "Can We Make OSHA and Workers' Compensation Insurance Interact More Effectively in Promoting Workplace Safety?" *Research in Labor Economics*, Vol. 10, 1989, pp. 1-52, © 1989, JAI Press Inc.; and grateful to the Cato Institute for permission to use material from Thomas J. Kniesner and John D. Leeth, "Improving Workplace Safety, Standards or Insurance?" *Regulation*, 14(4), Fall 1991, pp. 64-70, © 1991, Cato Institute.

Because the book is based on our previously peer reviewed published research simulating job safety policy we thank the numerous anonymous referees who contributed to making our try for a coherent framework as error free as possible. In addition, we appreciate the audience members, too numerous to mention, who have sat through our presentations at university seminars and professional conferences in the

United States, the Netherlands, Japan, and Australia, where we learned much concerning labor market institutions and government policies toward worker safety.

We have also benefited from time spent at Keio University's Economic Observatory, the Impact Centre of the University of Melbourne, the Department of Economics of LaTrobe University, the Center for Economic Research of Tilburg University, Stanford University's Hoover Institution, the Department of Economics at University College London, and the Department of Economics of the Research School of the Social Sciences and the Department of Statistics of the Faculties of the Australian National University.

Our book could not have been written without the generous financial support from many sources. We have been helped by a grant from the Workers' Compensation Project of the University of Connecticut via the Employment Standards Administration of the U.S. Department of Labor, who bears no responsibility for our policy recommendations. We have also been fortunate to have had sabbaticals from Bentley College and Indiana University, during which we wrote much of our book. Incidental manuscript preparation expenses and expenses for out-of-pocket research costs have also been generously provided by Bentley College and the Office of the Vice-President for Research and the University Graduate School of Indiana University.

Now for some names. Deborah Garlow gets our gratitude for checking references, improving the dialogue overall, and helping us have tables that make sense. Les Boden of the Workers' Compensation Research Institute gave us very insightful comments on Chapter 2. There are also two people we believe have contributed much to the development of the unified simulation approach we have used to examine labor market outcomes. Donald Cullen of Cornell University, as Editor of the *Industrial and Labor Relations Review*, which published our work on moral hazard problems in workers' compensation insurance, pushed us to think about the role of simulation in quantitative labor market research — specifically how numerical simulation complements econometric research. William Niskanen, as head of the Cato Institute, asked us to think about global reforms in workplace safety policy in the United States for a conference he organized on the government's role in risk regulation in all walks of life. Bill Niskanen's thoughtful prodding led to our final chapter, which benefited greatly from the experts at the Cato Conference, including Justice Stephen Breyer.

Our last set of kudos we give to two people whom we believe are the best in the world at what they do. Zachary Rolnik of Kluwer Academic Publishers deserves effusive praise as the best book editor both of us have ever had. Zachary is a patient problemsolver. There was no production glitch he could not help us overcome immediately. Zachary also understands and contributes to the topics of the books he edits. Anyone would be lucky to have Zachary edit a book. Finally, Kip Viscusi had the idea for us to pull together our previously published work, fill in the gaps in its links to policy and econometric research, and write the book that follows. It is rare to be able to say you know the leading expert on something, but we are grateful for the help all through our careers that we have gotten from Kip, who is the most knowledgeable person on the issues of risk and policy toward risk ever to have lived. Kip is a true giant because he does not see topics as his exclusive publications preserve but rather is a field builder and nurturer of his fellow researchers.

Thomas J. Kniesner

John D. Leeth

1 INTRODUCTION

North Carolina is best known for tobacco, but its biggest farm enterprise is poultry. More than 24,000 North Carolinians are involved in the growing, slaughtering, and processing of chickens (Smothers 1991). By all accounts working conditions in most poultry processing plants are horrendous. Plants are frequently hot and greasy, and the air reeks of chicken, ammonia, and chlorine. Many workers use dangerous and poorly maintained equipment at an extremely rapid pace with few breaks. The repetitive motions necessary to cut, cook, weigh, and package chicken parts produce a high incidence of hand and wrist injuries. According to Linda Cromer, director of organizing for the Retail, Wholesale and Department Store Union, poultry processing "is the worst industry in the nation" (May 1991). We introduce our unified numerical simulation of the U.S. safety policy system to follow with an explanation of how the government and the marketplace can permit jobs that are as dirty and dangerous as those in the chicken processing industry.

Conditions at the Imperial Food Products plant in Hamlet, North Carolina, may have been among the worst in the nation but nothing prepared the workers for the nightmare that began Tuesday morning,

September 3, 1991. At about 8:15 a hydraulic line for a conveyor belt carrying chicken parts ruptured. Normally the fluid in the line is not easily combustible, but when the line broke the fluid vaporized and was quickly ignited by the flames of a nearby gas fueled fryer. The jointly burning insulation, soybean oil, and chicken parts produced a thick yellowish smoke that quickly filled the sprawling windowless building of 30,000 square feet. When the fire made the electricity go out the inside of the building was soon dark, and the situation for the chicken processing workers became chaotic (Smothers 1991, Taylor 1991).

Attempting to escape the thick smoke, about 40 of the 90 workers in the building at the time rushed to the loading area but found the entry door locked from the outside. The group separated with some seeking refuge in a cooler and others fleeing to the loading dock. There was no exit from the cooler, and no escape from the loading dock was possible because it was blocked by a delivery truck and a trash compactor. Hearing shouts of fire, workers in another area of the plant rushed to the canteen but the door to the outside was padlocked. Eventually, employees kicked open the canteen door, moved the delivery truck, and unlocked the entry door in the loading area, but not before 25 people had died and 55 others had been injured, most from smoke inhalation (Taylor 1991).

Almost all the people who died or were hurt in the Imperial Food Products fire tried to escape through exits that were locked or blocked. Rescuers found three bodies at the loading dock. Most of the workers who fled from the loading area to the cooler died. Rescuers discovered 12 bodies in the cooler and three more just outside the cooler. The remaining seven workers were in another part of the plant; six workers were near the door of a freezer, and one worker was inside the freezer. No one knew whether the workers had originally attempted to escape through one of the blocked passageways before seeking refuge in the freezer (Associated Press 1991a).

The fire at Imperial Food Products was the worst industrial accident in North Carolina history. Outraged labor leaders, members of Congress, civil rights advocates, and others interviewed portrayed the loss of life as the needless result of laxly enforced health and safety regulations. During the 11 years the Hamlet, North Carolina, plant had been operating it had never been inspected by the Occupational Safety and Health Administration (OSHA). Workers claimed that the plant had been plagued by many small fires around the fryer and in the area where they collected refuse from the fryer. After an earlier blaze in 1983 the Hamlet Fire Department required Imperial Food Products to install a modern carbon dioxide based automatic extinguisher on the fryer's hood, but the mandated extinguisher system was useless in stopping the sequence of events leading to the 1991 disaster.

Although the owner of Imperial Food Products, Emmett Roe, should have known of the potential danger from fire he neither installed automatic heat detection sprinkler systems in the building nor designed a fire evacuation plan. Mr. Roe had purchased only one fire extinguisher for the entire building, and the fire doors did not meet national safety standards. Because he wanted to keep flies out of the plant and was also concerned workers were pilfering chickens, Mr. Roe had ordered that six of the nine doors to the outside be locked (Associated Press 1991a).

WHY OSHA DID NOT PREVENT THE TRAGIC FIRE

U.S. Representative William D. Ford, a major proponent of strengthening OSHA's enforcement powers, said soon after the fire that "If just one OSHA inspector — state or federal — had made Imperial Food Products unlock just one door, this tragedy might not have happened" (May 1991). Similar to 23 other states North Carolina is responsible for enforcing federal OSHA health and safety standards. At the time of the fire the North Carolina Labor Department had 27 safety inspectors and trainees to cover 180,000 workplaces (May 1991). With so few inspectors the state had not investigated 61 complaints of serious workplace health and safety hazards; a few uninvestigated complaints dated back to 1984 (Taylor 1991). The frequency of injuries and illnesses in poultry processing placed the Imperial Food Product's plant in the high hazard category, a designation requiring more frequent inspections. In 1990 North Carolina had inspected only 100 manufacturers for general safety violations and only five manufacturers for general health violations (*Wall Street Journal* 1991). According to Margaret Seminario, AFL–CIO director of occupational safety and health, a high hazard company in North Carolina could expect to see a state safety inspector once every 30 years.

State budget problems and the inability to attract candidates prevented North Carolina from hiring the 37 additional inspectors needed to meet federal staffing requirements. In the 1990 fiscal year North Carolina returned $453,000 in unspent federal funds to OSHA because the state would not meet the 50 percent matching requirement. Despite North Carolina's lackluster administration the year before the fire OSHA evaluated the state program as being effective (Taylor 1991). Reacting to the public outcry after the fire OSHA moved 14 federal inspectors to North Carolina to investigate worker complaints of safety and health violations. For the first time in its history OSHA took over part of a state's responsibility for enforcing health and safety regulations (Associated Press 1991b). OSHA also set up a toll free nationwide hot line for employees to report without fear of retaliation unsafe working conditions (1-800-321-OSHA).

Had OSHA stepped in earlier could it have prevented the tragic fire at the chicken processing plant in Hamlet, North Carolina? By most accounts OSHA probably could not have prevented the tragedy. Imperial Food Products stood only a slightly higher chance of being inspected by OSHA overall than by the North Carolina Labor Department. In 1991 there were 1,186 federal safety and health inspectors to protect 117 million workers (Tye 1991). Based on the 1991 inspection rate an average American worker can expect to see an OSHA inspector once ever 75 years or once every 13 years if working in a hazardous job (Bernstein 1991).

Even if OSHA had discovered the serious fire hazards at Imperial Food Products it still might not have forced compliance soon enough to prevent disaster. As shown by its treatment of the Pymm Thermometer Company, the second largest manufacturer of mercury thermometers in the United States, OSHA inspectors are often reluctant to close a company down or even impose dramatic fines when they find serious violations of health and safety standards.

Employees at Pymm worked without respirators or protective clothing in a building lacking nearly any outside ventilation. The worst conditions were in a windowless room in the cellar where workers recycled mercury from old thermometers. Exposure to mercury can cause permanent damage to the brain, lungs, liver, and kidneys. The basement mercury recycling room was filled with broken glass and noxious fumes from puddles of mercury on the floor (McGarity and Shapiro 1993, p. 211). In 1987 more than half the Pymm Thermometer Company's 100 employees and some of their children showed signs of mercury poisoning. One Pymm worker had suffered permanent brain damage.

Until a former employee tipped off health and safety inspectors OSHA was unaware of the mercury recycling room, but for five years the agency had known of other serious health problems at Pymm. OSHA first cited the company in January 1981 for exposing workers to mercury fumes five times greater than normal, a level capable of causing serious neurological damage or death. OSHA fined Pymm $1,400 and ordered it to eliminate the hazards within six months. By five years after the original elimination order, during which there were three additional inspections, Pymm had still not installed the ventilation equipment needed to reduce mercury exposure to permissible levels.

The situation at Pymm Thermometer may have continued unabated except for the adverse publicity caused by the firing of Vidal Rodriguez. Doctors reported to the New York City and the New York State health departments that Mr. Rodriguez was suffering from mercury poisoning after he was treated for a fractured elbow. Afflicted with asthma and no longer able to walk without a cane, Rodriguez applied for workers' compensation benefits and was promptly fired by Pymm. Newspapers carried accounts of Mr. Rodriguez's case, including concerns of the New York City and State Health Departments over working conditions at the thermometer company. OSHA still did nothing. Finally, an article detailing OSHA's failed record of enforcing health standards goaded the agency into reinspecting Pymm in October 1985 (Bennett 1991).

Mr. Rodriguez, who lived only one block from the thermometer plant, stopped the safety inspectors on the street and told them of the recycling operation in the basement. The inspectors searched for more than a day to find the hidden room and, although they said working conditions were "a nightmare," they fined Pymm only $30,100 for 16 violations of federal workplace safety and health standards (Noble 1986). A year later, two months after the state of New York indicted Pymm's owners for assault for exposing their workers to high levels of mercury, OSHA inspected the plant again and issued another $75,000 in citations (Roel 1988). Even with federal administration, health and safety inspectors failed to uncover deadly working conditions at Pymm and let stand serious health threats for more than five years.

Other case histories show that the situation at the Pymm Thermometer Company was far from unique. OSHA inspectors frequently overlook dangerous working conditions and, even when they find serious health and safety violations, inspectors often cannot compel companies to eliminate the hazards. To encourage timely compliance administrators frequently slash assessed penalties, further reducing the already minor economic incentives for firms to observe health and safety standards at all. Firms realize, in the unlikely event they are inspected, they can avoid paying significant fines by simply agreeing to abide by OSHA's regulations in the future (Lofgren 1989).

U.S. WORKPLACE INJURIES AND ILLNESSES

With an annual budget of about $300 million, OSHA is about five percent the size of the Environmental Protection Agency (Salwen 1993). The federal government has six times more fish and game inspectors than health and safety inspectors (Ball 1992). As demonstrated in the Imperial Food Product's fire, OSHA frequently steps in only after a disaster. One might believe the limited federal commitment to worker health and safety would lead to a vast and expanding frequency of workplace accidents and diseases. As shown in Figure 1.1 the frequency of workplace deaths as measured by National Safety Council statistics has declined dramatically over the last 64 years. In 1928, for every 100,000 workers there were 15.8 workplace fatalities. By 1992 the rate of fatalities had fallen by about 80 percent to 3.3 per

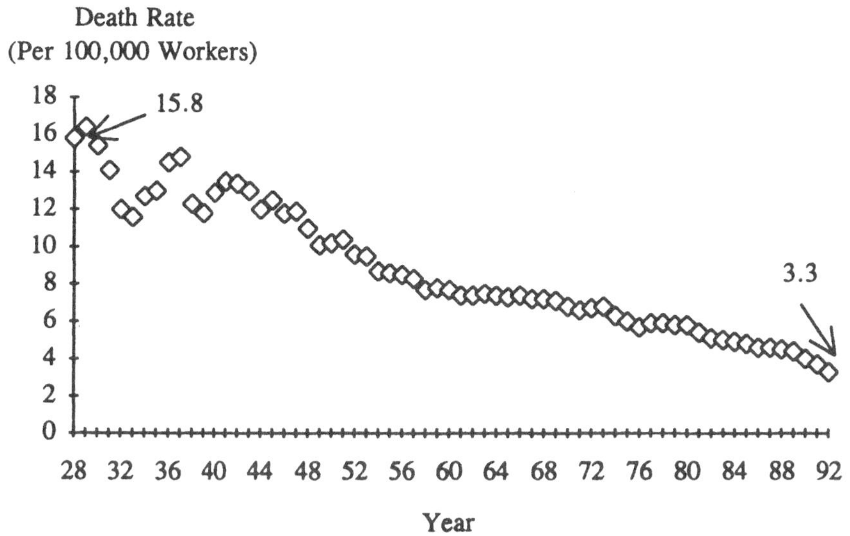

Figure 1.1 Workplace Fatalities, 1928–1992

100,000 workers. By comparison, in 1992 the chance of dying in a motor vehicle accident was about five times greater (15.8/100,000) and the chance of dying in an accident at home was over two times greater (7.6/100,000) than the chance of dying at work (National Safety Council 1993).

Because of the estimating procedure the National Safety Council statistics may undercount U.S. workplace fatalities. The Bureau of Labor Statistics (BLS) has also calculated workplace fatality rates since the late 1930s but the scope of their survey has changed so dramatically that one cannot use the BLS data to identify a time trend. Nevertheless, the BLS's most recent data collection effort creates the most reliable source of information on fatal workplace accidents in the United States.[1] The BLS conducted a census of fatal occupational injuries for 1992 using data from death certificates, workers' compensation claims, medical examiners' records, autopsy reports, motor vehicle accident records, and OSHA and Mine Safety and Health Administration fatal injury reports. The BLS estimated the average worker in the United

States faced a 5 in 100,000 chance of dying on the job in 1992. Although higher than the National Safety Council estimate the BLS figure of a 5/100,000 fatality rate still implies workers faced greater hazards at home than at work. More interestingly, the BLS also found that 40 percent of all workplace fatalities were caused by transportation accidents (45 percent of the fatal transportation accidents were highway accidents), and 20 percent of workplace fatalities were caused by assaults and other violent acts (85 percent homicides and 15 percent suicides). Neither transportation accidents nor assaults are likely to be reduced much by OSHA inspections for workplace health and safety violations.

Data on nonfatal workplace injuries and illnesses paint a somewhat different picture than fatal injuries. Figure 1.2 shows the trend of nonfatal workplace injuries and illness since 1973, which is the first year firms were required to report industrial accidents and diseases. Unlike death rates, injuries and illnesses do not show a marked decline over time. Since 1975 the pattern of workplace injuries follows closely

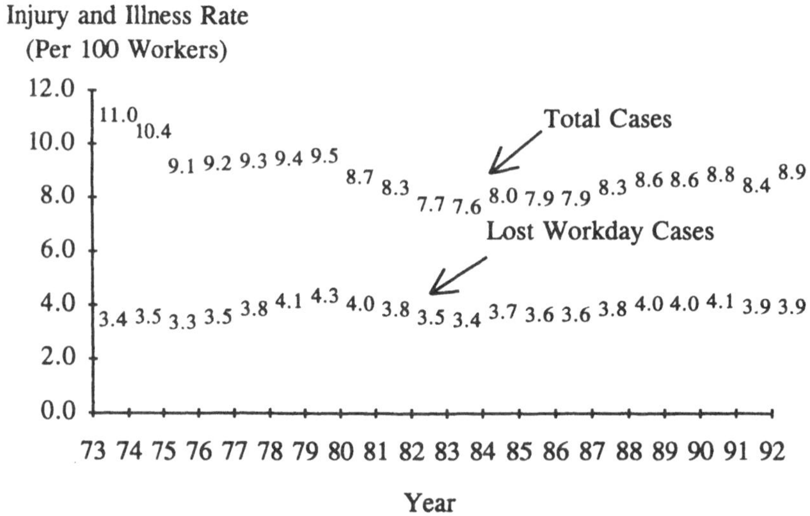

Figure 1.2 Workplace Injuries and Illnesses, 1973–1992

the business cycle. Injuries and illnesses rose slightly during the business upturn of the late 1970s, fell during the recession of the early 1980s, and then rose again during the 1980s macroeconomic expansion.

Arguments for expanding OSHA usually begin by noting the high rate of nonfatal workplace injuries and illnesses in the United States. For every 100 full time equivalent workers in 1992 there were 8.9 reported cases of nonfatal accidents and diseases. The majority of injuries were relatively minor, neither limiting work activities nor requiring lost worktime recuperation. Still, one should neither minimize the suffering of the workers involved nor ignore the high fraction of relatively severe injuries. At a rate of 3.9 per 100 workers, lost workday injuries and illnesses were 44 percent of all nonfatal workplace accidents and diseases in 1992. U.S. firms reported almost three million nonfatal disabling injuries during the year (BLS 1993).

The cost of workplace injuries is enormous. In 1992 the U.S. economy lost $115.9 billion from work-related accidents: $62.5 billion from wage and productivity losses, $22.0 billion from medical costs, $14.5 billion from administration expenses, $3.4 billion from motor vehicle damages, $10.2 billion from indirect employer costs, and $3.3 billion from fire losses (National Safety Council 1993). In addition to monetary losses, work injuries cause pain and suffering and frequently result in permanent disabilities that impede the normal enjoyment of life. The National Safety Council (1993) estimated that for every dollar of monetary loss accidents lower the quality of life on average by two dollars. With a two-to-one quality-of-life loss ratio the total cost of workplace accidents in 1992 was $347.7 billion, about 5.8 percent of 1992 U.S. GDP.

WORKPLACE SAFETY PROGRAMS IN THE UNITED STATES

More than the economic costs of workplace accidents and diseases the human suffering of the victims and their families drives the desire to design government programs to reduce the frequency and the severity of work related injuries. Until the early part of the twentieth century an injured worker had to prove employer negligence in court to collect damages, either medical or compensatory. Firms could avoid liability if they could show in court that the injury was the worker's fault, was an ordinary hazard of employment, or was caused by a coworker's carelessness. Under the legal system for determining fault prior to workers' compensation insurance (WC) people injured at work could neither count on receiving compensation nor be sure of the amount or the timing of injury compensation payments. Many believed the pure negligence system with its uncertain liability payments and generous escape clauses provided little protection for workers and offered firms few incentives to improve workplace safety.

Workers' Compensation Insurance

Passage of workers' compensation insurance laws made firms strictly liable for industrial injuries. Regardless of fault employers must pay all of an injured worker's medical bills and a portion of his or her lost income. Firms provide WC coverage either by purchasing insurance from a public or a private carrier or by self insuring after proving financial competency. In return for no fault insurance coverage workers cannot sue employers for damages due to work related injuries. Firms can lower their costs of purchasing WC coverage by reducing the frequency and severity of workplace accidents. Proponents of WC believed the movement to a strict liability system under WC would shift accident costs to firms and encourage greater safety efforts.

The Occupational Safety and Health Act of 1970

By 1970 many policymakers felt the WC system had failed to keep its promise of promoting a safer workplace. From 1958 to 1970 the average manufacturing injury rate grew by about two percent per year (Viscusi 1983). The unexpected rise in injuries motivated passage of the Occupational Safety and Health Act of 1970, the federal law empowering OSHA to establish and enforce mandatory safety standards.

The Act established three violation categories: (1) nonserious, (2) serious, and (3) willful and repeated. The law does not require OSHA inspectors to penalize companies for nonserious violations of health and safety standards. Inspectors must, however, levy fines for serious violations, which are infractions substantially increasing the likelihood of a worker fatality or serious injury, and to levy fines for willful and repeated violations. Under the original Act inspectors could fine firms up to $1,000 for each serious violation of health and safety standards, $10,000 for each willful violation, and $1,000 per day for each repeated violation. In 1990 Congress increased maximum fines to $7,000 for serious violations and increased maximum fines to $70,000 for willful and repeated violations.[2] The Occupational Safety and Health Act established criminal penalties only in situations where willful violations result in worker deaths.[3]

Many argued that OSHA enforcement of health and safety standards would dramatically reduce work hazards. Firms would no longer merely be enticed to improve safety through the possibility of lower WC insurance premiums; OSHA would force firms to provide a safe working environment. Despite high hopes OSHA has only marginally reduced work-related injuries. The difference between WC and OSHA is not that one program entices and the other forces; both rely on financial incentives to encourage workplace safety. With WC, firms reduce work hazards because a drop in injuries and, therefore claims for WC benefits, ultimately lowers the cost of purchasing WC insurance. With OSHA, firms reduce work hazards because obeying OSHA's dictates lowers the expected monetary penalty for noncompliance.

Firms observe OSHA standards for the same reason drivers observe speed limits; they fear they will be ticketed and fined. More employers will obey safety standards and more drivers will obey speed limits if the chance of being caught is fairly high. Likewise, holding the probability of detection constant, more firms will comply with OSHA regulations and more drivers will obey speed limits if fines for violations are greater. Employers, like drivers, react not to the existence of the law but to the enforcement of the law as measured by the likelihood of detection and the penalty for noncompliance. OSHA can expand compliance by stepping up its enforcement efforts; it can increase the number of firms inspected, raise the number of violations per inspection, or boost the average fine per violation. Firms install additional safety equipment and establish new safety programs because OSHA's enforcement efforts raise the financial benefits of reducing workplace hazards.

The Comparative Advantages of WC and OSHA

Both WC and OSHA rely on financial incentives to encourage workplace safety. The approach of OSHA to encouraging workplace safety is very different from how WC tries to encourage workplace safety.

OSHA standards have generally required firms to install specific types of safety equipment or to initiate specific safety programs. The regulatory approach assumes a government agency specializing in workplace safety will know more about how to make a work site safer than employers and workers. Opponents argue OSHA's safety measures have frequently been ineffective or inefficient; standards either do not reduce hazards or they reduce hazards at a greater expense than other methods available.

In contrast, WC leaves the design of safety programs up to employers. The mandatory no fault insurance approach assumes employers and workers are best able to determine appropriate safety measures given the individual circumstances of their working environments. Critics argue WC does not adequately prevent employers from exposing workers to extreme health and safety hazards because for some employers the cost of additional safety programs may outweigh the benefit of lower WC insurance premiums.

The financial beneficiaries of WC and OSHA also differ. With WC the insurance premiums help compensate either injured workers or their families, whereas with OSHA the revenues collected help defray other government expenses.

Finally, both OSHA and WC influence employer behavior, but only WC influences worker behavior. Insurance benefits reduce the economic hardship from industrial accidents and diseases and may directly encourage workers to be less concerned with job safety in the first place. The urge to receive or extend insurance benefits may also entice some workers to report bogus or exaggerated disability claims under WC.

ECONOMETRICALLY STUDYING JOB SAFETY

In the next chapter we survey the econometric evidence concerning how federal OSHA regulation of the workplace and states' workers' compensation insurance laws have affected U.S. worker safety. As background for our discussion of econometric research we believe it valuable to provide some discussion of issues to keep in mind when considering the econometric literature we survey in Chapter 2 as a launching pad for developing our numerical simulation model in Chapter 3.

The mode econometric model of job safety has been based on a reduced form equation such as

$$\pi_j = f(s_j, x_j) + \varepsilon_j, \tag{1.1}$$

where a measure of the rate at which workers are injured (π_j) is the dependent variable, measures of safety programs (s_j) and the economic environment conditioning labor market outcomes (x_j) are the independent variables, and the appended error term (ε_j) captures unmeasured (ideally) randomly occurring factors influencing injuries. The subscript j either indexes time periods, such as years before versus after the passage of OSHA or WC, or j indexes states, whose WC laws differ in benefit generosity and insurance premium pricing. Questions of measurement are basic in evaluating the quality of econometric research on the safety effects of OSHA or WC.

The first question to consider in an econometric study is the accuracy and breath of the dependent variable. Is the outcome being studied all injuries at work or only fatalities? Fatalities are less common but measured more accurately than nonfatal injuries. Second, are only injuries included or are diseases and possibly subtle effects of the job on health, such as stress, measured too? Moreover, what is the basis for measuring the dependent variable? Do injuries come from a sample of workers or an actual census of health records, say? On the other side of equation (1.1) one need pay attention to how the program effects are captured by s_j. Are program effects inferred simply from injury rates before and after the introduction of OSHA or WC or does the regression contain actual program parameters, such as OSHA inspection rates and fines or interstate differences in WC benefit generosity and the degree to which insurance premiums reflect firms' injury histories? Finally, when evaluating econometric research a reader needs to consider the details the study provides about safety effects. Does the regression estimate only qualitative inferences about the programs' effects or does the model yield quantitative safety effects in enough detail to infer a program's cost effectiveness? Having emphasized the importance of how safety outcomes and program effects are measured we now dig more deeply into the complexities of econometric models of safety in the workplace.

The Canonical Hedonic Labor Market Model.

For a more complete picture of how safety policy affects labor market outcomes, including worker well-being, a researcher needs to examine the structural equations that underlie the popular regression equation summarized by (1.1). The canonical structural econometric model of the labor market where a sorting of workers into jobs yields a set of wage and safety levels for each job and worker, known as hedonic labor market equilibrium, has the three stochastic equations

$$w_j = W(\pi_j, x_j^L) + \varepsilon_j^L, \tag{1.2}$$

$$n_j^s = N^s(W'(\pi_j), x_j^s; \alpha_j) + \varepsilon_j^s, \text{ and} \tag{1.3}$$

$$n_j^d = N^d(W'(\pi_j), x_j^d; \mu_j) + \varepsilon_j^d. \tag{1.4}$$

The first equation in the system (1.2) is the hedonic equilibrium locus, summarizing how the wage rate indexed by j (which could indicate a person, industry, occupation, location, or year) varies with the injury rate and other characteristics that influence wage outcomes in the labor market (x^L). Equations (1.3) and (1.4) are the supply and demand for workers that produce the hedonic locus in (1.2). The supply and demand for workers both depend on three things: (1) the marginal wage ($W'(\pi_j) \equiv \partial W(\pi_j) / \partial \pi_j$), which varies with the injury risk, (2) the vectors of usual independent variables influencing the number of workers and jobs captured in x^d and x^d, such as households' wealth and firms' nonlabor input costs, and (3) workers' attitudes toward risk as metered in the single person specific random variable α plus firms' efficiency in producing safety as summarized in the single firm specific random variable μ. Estimating the supply schedule in (1.3) identifies the parameters of workers' utility functions, and estimating the demand relation in (1.4) identifies the parameters of the firms' production and cost functions. Each equation in (1.2)–(1.4) plays a different role in determining workplace safety, and each equation has special statistical properties that make (1.2)–(1.4) differ from the usual simultaneous equations system, which makes estimation particularly difficult and creates a role for numerical simulation as a complement to econometric research.

Estimating the Hedonic Wage Locus. Economists use the hedonic locus summarized in (1.2) to study employment patterns of how the labor market implicitly compensates workers for workplace injury risks. The slope of the hedonic locus, $W'(\pi_j)$, is the marginal wage premium firms must pay workers to accept (more) injury risk and has been used to compute the implicit value workers place on their lives via risking death at work (Viscusi 1992, 1993).

A special difficulty in estimating the hedonic equilibrium wage function in (1.2) from data on individual worker choices comes from unmeasured differences across workers in wealth or labor market productivity (Hwang, Reed, and Hubbard 1992). In particular, the error term in (1.2), ε_j^L, may not be random. Because it captures all influences not measured in the independent variables ε_j^L may vary across workers because of unmeasured productivity or wealth differences that help to determine the worker's choice of where to work, and in turn, determine the risk of injury. Because more productive or wealthier workers generally want greater safety (lower π_j jobs) the econometric consequences of unmeasured worker wealth or worker productivity) is a biased estimate of the slope of the hedonic locus, $W'(\pi_j)$.

Computations using reasonable assumptions of unobserved worker productivity differences suggest that the true slope of the hedonic locus can be as much as 10 times greater than the typical econometrically estimated slope (Hwang, Reed, and Hubbard 1992). The extreme downward bias in the estimated compensation workers receive for accepting injury risk is problematic for cost effectiveness studies of policy because it produces too low a value on job safety and workers' lives, which are crucial numbers when evaluating OSHA and WC. Complex econometric techniques requiring longitudinal data (a set of workers surveyed for multiple time periods) or better data on worker wealth or productivity would improve the accuracy of econometric estimates of workers' compensation for risk. Because more complete and accurate data may not soon be available our book explores the value of the numerical simulation alternative for drawing quantitative inferences concerning the compensation for risk as summarized in the hedonic wage function (1.2).

Our approach in Chapters 3–5 is to begin with a hedonic wage function that mimics the situation in U.S. labor markets prior to OSHA (Thaler and Rosen 1975). The key to proper use of numerical simulation is to begin with a set of parameters that reproduces the outcomes of interest (Watson 1993) and then examine the robustness of the behavioral implications to less than perfect knowledge of the underlying behavioral relations. In using numerical simulation to study the quantitative properties of hedonic equilibrium and how it is disturbed by changes in safety regulation and injury insurance we are careful also to see if the starting values we use in our simulations make much difference to the policy implications.

Estimating the Workers' and Firms' Value Functions. For certain policy questions it is necessary to identify the objectives or so-called value functions that underlie the parameters of the supply and demand for workers in (1.3) and (1.4). For example, if we want to draw inferences concerning how workers' well-being or firms' profits change due to more stringent OSHA regulations then we will need the parameters of utility functions and profit functions. There is an extensive theoretical econometric literature explaining why identifying the parameters of the structural equations in the canonical hedonic model in (1.2)–(1.4) is more difficult than the standard simultaneous equations system, such as the supply and demand for a homogenous product like milk (Brown and Rosen 1982, Brown 1983, Epple 1987, Kahn and Lang 1988). The primary reason that identifying the supply and demand equations underlying hedonic equilibrium models is so econometrically difficult is that the prices are implicit, rather than posted.

While you can go to the grocery store or look in the newspaper to learn the price of eggs, the price of extra workplace safety is implicit in the wage structure. The hedonic wage equation (1.2) not only describes the labor market in equilibrium but its parameter estimates are also what the econometrician must use to compute the implicit price of workplace safety, $W'(\pi_j)$, which is then used as an independent variable in the supply and demand schedules (1.3) and (1.4). A crucial consequence of having to manufacture econometrically the implicit price workers pay for (the implicit costs to firms of providing) workplace safety is that either $W(\pi_j, x_j^L)$ in (1.2) must be highly nonlinear (for example, cubic in π) or some of the independent variables explaining the hedonic locus in x_j^L, such as region, cannot also explain either labor supply or demand (x_j^L must differ from x_j^s and x_j^d). The logic behind the restriction on the shape of and explanatory variables in (1.2) is that the hedonic wage function must contain information independent of supply and demand if the implied compensating wage differential $W'(\pi_j)$ computed from estimates of (1.2) is to add additional information to the other arguments in supply and demand (1.3) and (1.4) (Brown and Rosen 1982).

Another econometric complexity in estimating a complete hedonic equilibrium model as summarized in (1.2)–(1.4) happens because the rate of workplace injuries (safety), π, is the result of optimizing decisions by firms and workers. The observed injury rate must be treated statistically as an endogenous variable. If we were in the standard simultaneous supply and demand econometric model the researcher could replace all endogenous variables on the right-hand side of an equation with values predicted from a reduced form equation, which is a regression of each endogenous variable on all exogenous variables in the system. How the researcher addresses econometrically the dual endogeneity of pay and workplace safety as workers and firms match in hedonic labor market equilibrium is more restricted than in the supply and demand for homogeneous commodities.

A special econometric complexity is that workers with unusually high desires for workplace safety (low πs) given their latent attributes (e^s) will match with firms that are low cost producers of safety as captured in x_j^d. Similarly, workers unconcerned with job safety will tend to match with firms where safety is difficult and expensive to produce. The result is that ε^s and x^d are not statistically independent; likewise ε^d and x^s are not statistically independent. The practical econometric implication of how workers and firms match is that in predicting (also so known as instrumenting) the values of regressors in the supply schedule the researcher may not be able to use exogenous variables from the demand schedule, and in predicting the values of regressors in the demand schedule the researcher may not be able to use exogenous variables from the supply schedule. Thus, there is limited information available to identify the parameters of the workers' preferences embedded in their supply equations (1.2) and limited information available to identify the parameters of firms' costs and technologies embedded in their demand equations (1.3), which complicates inferring how policy affects economic well-being.

Enter Numerical Simulation.

Because of the extreme data demands of a complete structural hedonic model of the labor market it is convenient to consider the alternative of numerical simulation. Instead of trying to use data to identify econometrically firms' and workers' value functions we take the research in another direction via numerical simulation.

We begin with what we know from the econometric literature concerning workers' heterogeneous preferences toward risk and firms' heterogeneous costs of making the workplace safer and formulate a computable hedonic model similar to (1.2)–(1.4). In particular, we solve numerically for a set of outcomes that replicates the pre-OSHA labor market in the United States. We then introduce OSHA plus refinements to workers' compensation insurance into the model to compute the new equilibrium that results from workers and firms reoptimizing in the face of a change in the policy regime. A before and after numerical comparison then allows us to infer the safety and welfare effects of public policy intended to make workplaces safer and workers economically better off. At every step we are careful to check the robustness of our results to possibly imprecise parameters of the economic relationships driving the numerical simulation model.

We do our numerical simulations with three objectives in mind. We want to discover any subtle complementarities between the safety effects of OSHA and WC, to distinguish among the various incentives that WC offers to reduce injuries at work versus report possibly false injuries for insurance payments, and to discover possible cost effective reforms to OSHA and WC as they currently exist in the United States.

THE SIMULATION APPROACH TO STUDYING JOB SAFETY

Many policy analysts have recommended dramatic increases in OSHA's enforcement powers (Lofgren 1989, McGarity and Shapiro 1993). They call for more inspections, larger fines for violations, better training of inspectors, and an expedited process for designing and implementing safety and health standards. Having generally failed to improve workplace safety OSHA should now do more of the same, only better. The question is how much will safety improve if we expand OSHA's enforcement powers? What are the quantitative links between enforcing workplace safety standards and safety outcomes?

Our book's focus is on examining quantitatively the ingredients of workplace safety policy in the United States with the goal of identifying refinements in current programs to make them more cost effective in promoting a safe working environment. Our approach is unique. Building on recent research in applied general equilibrium we developed a computable hedonic labor market model in which workplaces vary by level of safety and workers vary in their willingness to accept injury risk. Numerical simulation is necessary because it is difficult to identify econometrically the complete economic structure underlying labor market outcomes needed to make quantitative comparative policy evaluations, and existing data are insufficient to infer the effects of some of the extreme changes in OSHA and WC that we consider.

Figure 1.3 presents an overview of the numerical model we develop and use to guide our discussion of the economics of safety in the workplace. Our integrated approach to the linkages among the labor, product, and insurance markets as tempered by government intervention focuses on how changes in safety policy alter labor market outcomes. Unlike other treatments we realistically allow workers to have differing views about risk and firms to have differing capabilities to produce a safe working environment. Firms unwilling or unable to eliminate all job hazards must pay higher wages to attract workers away from firms offering complete safety. The relationship between wages and risk, the so-called hedonic wage function, adjusts to match workers to jobs so as to equalize the supply of workers with the demand for their services along the entire risk spectrum.

Changes in safety policy alter workers' and firms' decisions regarding workplace hazards, disequilibrating the labor market. The resulting change in wages, in turn, creates feedback effects that either reinforce or counteract the direct effects of government programs encouraging safety. Feedback effects occur both within the labor market and between the labor market and other institutions such as the market for disability insurance. Safety decisions influence the number of WC claims and ultimately the price of WC premiums. Likewise, safety decisions influence production costs and therefore the prices of goods and services. To capture the full set of subtle and not so subtle counterbalancing effects from government policy intended to increase work-

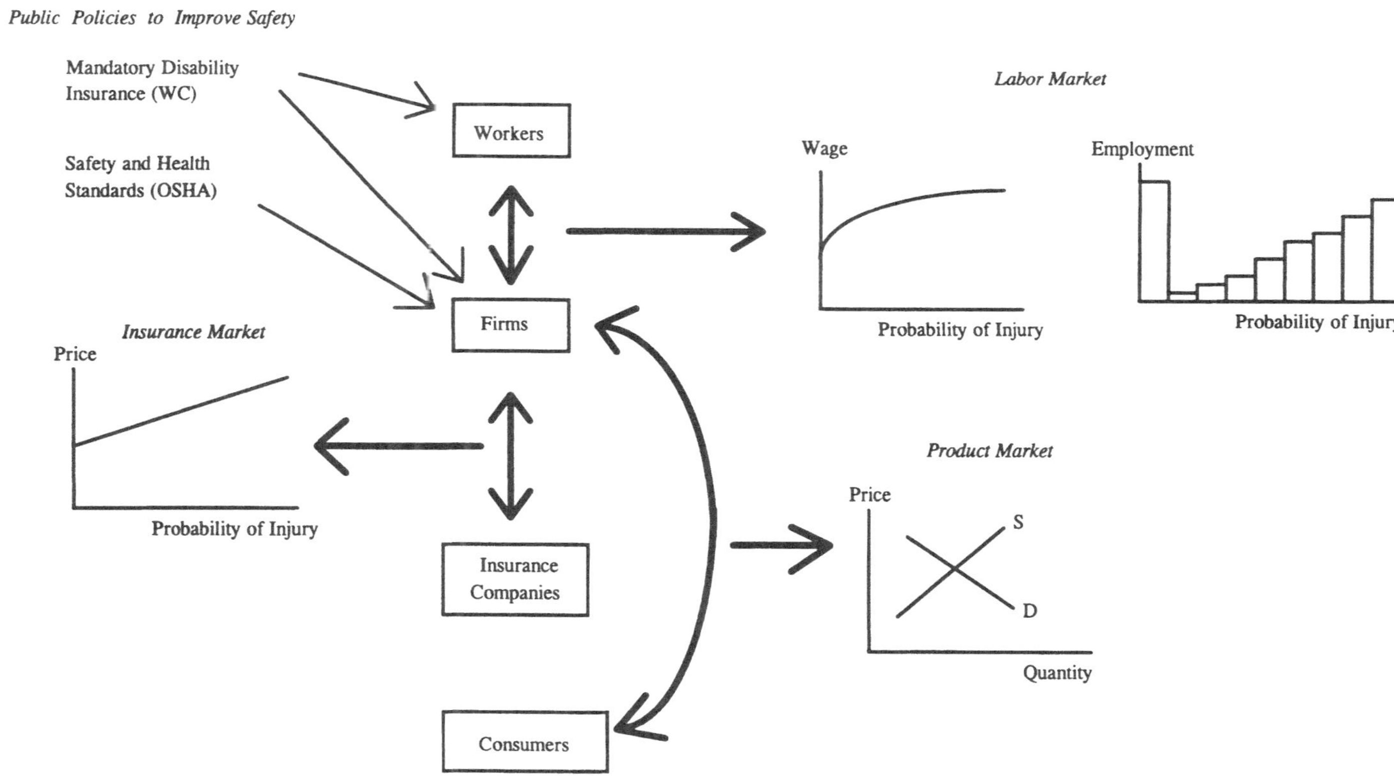

Figure 1.3 Safety Policy Interactions

place safety, our numerical model examines the complete set of interactions among the labor, insurance, and product markets using what economists currently know quantitatively.

SUMMARY OF KEY SIMULATION RESULTS

Our numerical simulations lead us to conclude that the process of matching workers to jobs in the labor market creates powerful incentives for an economically efficient level of workplace safety. We emphasize that efficiency does not mean zero risk. People will still get killed or injured on the job, just as they will die from or be hurt by accidents happening at home. Our numerical simulations illustrate that government policy with the objective of achieving greater workplace safety efficiently should move away from the regulatory approach that OSHA has used. On the other hand our calculations demonstrate that the experience rating characterizing WC insurance premiums is economically effective at encouraging workplace safety, and the link between a firm's safety history and WC insurance premiums could be further strengthened. Additional safety incentives could be introduced by permitting workers to sue negligent employers for pain and suffering from industrial injuries, which would make firms more economically aware of the true costs of workplace injuries. Because of the heterogeneity of workers and firms the best government policy seems to be to give maximal economic incentives for workplace safety while allowing firms and workers the freedom to discover how best to improve workplace safety on their own.

NOTES

[1]Because the ultimate cause of an illness is difficult to determine the BLS excluded diseases in their count of workplace fatalities (Toscano and Windau 1993).

[2]Correcting for the effects of inflation since 1970 the increase slightly more than doubled potential penalties for infractions of health and safety standards.

[3]From 1970 to 1988 OSHA had referred only 40 cases to the U.S. Justice Department for criminal prosecution, and the Justice Department had successfully prosecuted only two cases (Roel 1988).

2 THE ECONOMICS OF WORKPLACE SAFETY

Firms differ in their abilities to reduce or eliminate work-related health hazards. In 1992 workers in construction were almost five times more likely to be injured on the job and seven times more likely to die from an industrial accident than workers in finance, insurance, and real estate.[1] Additionally, within each industry some firms exhibit sterling safety records while others consistently have an exorbitant number of injuries. The extreme expense of removing or reducing many work-related health hazards prevents most companies from providing completely safe work environments. After comparing the costs of additional safety to the benefits from reduced injury rates, which include a lower wage premium necessary to attract workers from comparatively safer employment, many firms spend far less on safety than necessary to eliminate all hazards. On the other side of the market, workers must decide whether the additional wages received from relatively high risk jobs compensate for the greater likelihood of injury. Differences in attitudes regarding risk cause some workers to view a given wage premium as sufficient to outweigh the greater chance of an accident, while others view the wage premium as insufficient compensation.

Because employers differ in their abilities to reduce injury risk and workers differ in their attitudes regarding such risk, no single probability of a work-related injury is optimal for all parties. Here we describe the process generating labor market equilibrium when workers have differing views about risk and firms have differing capacities to produce safe work environments.

After developing the hedonic equilibrium framework we examine the safety impacts of the states' workers' compensation (WC) insurance and the federal government's Occupational Safety and Health Administration (OSHA) health and safety standards, the two major public policy programs intended to reduce industrial accidents and diseases in the United States. Finally, we review the empirical literature investigating the effectiveness of the two programs. Based on the best available information, OSHA has lowered workplace injuries by a fairly small amount, 5 percent at best. In contrast, research correctly controlling for worker incentives to report more injuries as WC benefits rise shows WC dramatically improving safety. The most optimistic study finds WC reducing workplace fatalities by 48 percent.

LABOR MARKET EQUILIBRIUM WITH WORK-RELATED HEALTH HAZARDS

In the standard labor market model all workers and firms are identical and a single wage equilibrates the quantity supplied of labor with the quantity demanded. If the wage is too low then the excess demand for workers drives the wage up, whereas if the wage is too high the excess supply of workers drives the wage down. When workers and firms differ in their attitudes regarding safety the process generating labor market equilibrium is only slightly more complicated. Instead of single wage, a wage function equilibrates the quantity of labor supplied to the quantity demanded at or near every possible level of injury risk (Rosen 1974, 1986).

Economists refer to the equilibrium relationship between wages and injury risk as a hedonic wage function. Wages reflect not only the overall conditions in the labor market but also the relative attractiveness (pleasure) of one job versus another. The underlying force generating the hedonic wage function is the sorting of workers and firms among the various levels of injury risks (safety). To simplify our discussion we assume all employers offer identical hours of work, and workers accept employment in only one firm. We also ignore the possibility of a worker individually purchasing WC or any other type of disability insurance. Finally, we will ignore for now variation in accident severity. Workers base their employment decisions on the probability of a work-related injury resulting in total incapacity or death.

Workers

The problem confronting a worker is to find the level of injury risk (π) that maximizes expected utility. Given our initial assumptions we can mathematically represent expected utility as

$$\bar{u} = (1-\pi)U(W(\pi)+y;\alpha) + \pi\tilde{U}(b+y;\alpha), \qquad \textbf{(2.1)}$$

where

$\bar{u}$	$\equiv$	the expected utility index,
π	$\equiv$	the probability of a work-related injury,
$U(\cdot)$	$\equiv$	the worker's utility function if uninjured, $U' > 0$ and $U'' < 0$,
$\tilde{U}(\cdot)$	$\equiv$	the worker's utility function if injured, $U > \tilde{U}$ and $U' > \tilde{U}'$,
$W(\pi)$	$\equiv$	the market wage function, observable to workers and firms, with $W' > 0$,
y	$\equiv$	nonlabor income,
b	$\equiv$	insurance benefits if injured, and
α	$\equiv$	a parameter determining an individual's aversion to risk.

Notice that expected utility is a weighted average of the utility if uninjured and the utility if injured with the weights equaling the probabilities of being (un)injured. Also notice the expected utility function explicitly considers both monetary and nonmonetary losses from workplace injuries. The difference between $U(\cdot)$ and $\tilde{U}(\cdot)$, income held constant, represents the pain and suffering resulting from an injury. Finally, $W(\pi)$ need not yet be an equilibrium wage function. Workers and firms need only observe and make their decisions based on the relationship between wages and injury risk.

By differentiating equation (2.1) with respect to π, setting the result equal to 0, and then rearranging we can show that a worker's optimal level of risk (safety) is when

$$U'\frac{\partial W}{\partial \pi} = U\big(W(\pi) + y;\alpha\big) - \tilde{U}(b + y;\alpha). \qquad \textbf{(2.2)}$$

The story here is the standard one — a worker weighs the marginal benefit of increased risk against the marginal cost. The left-hand side of equation (2.2) represents the marginal benefit, the added pay from a higher risk job, while the right-hand side represents the marginal cost, the increased likelihood of an injury that lowers both income and the utility from income. Because people differ in their degrees of risk aversion the perceived marginal gain and cost differ among people, in turn causing the optimal level of safety to vary. Workers with low risk aversion sort into high risk jobs, and workers with high risk aversion sort into low risk jobs.

A worker's acceptance wage function (indifference curve) shows the tradeoff between wages and risk at a constant level of expected utility. To maintain a specific level of expected utility wages must rise to compensate for bearing additional workplace hazards; acceptance wage functions slope upward. Additionally, workers more averse to risk require greater wage compensation for a given decrease in workplace safety than workers less averse to risk, all else equal; the worker with the steeper acceptance wage function at a given wage and level of safety is the more risk averse. Lastly, workers prefer higher wages to lower wages at any level of risk; the higher the acceptance wage function the higher the expected level of utility. Graphically the choice of the optimal level of safety can be viewed similarly to the choice of the optimal purchase of commodities with the market wage function replacing the standard income constraint.

Figure 2.1 portrays acceptance wage functions for two workers, Andrew and Barbara, in relation to a market wage function. We see Andrew maximizing expected utility by selecting a job offering a probability of injury equal to π^A. The highest level of utility he can achieve occurs where the acceptance wage function is just tangent to the market wage curve. Although π^A maximizes Andrew's utility, it does not maximize Barbara's utility who requires a smaller increase in wages to accept a slight rise in workplace hazards, expected utility held constant. She maximizes expected utility by choosing slightly more risk, π^B, and earning a higher wage. Given a sufficiently large number of diverse workers, each point on the hedonic wage function is a point of tangency for some group of workers. Technically, the wage function represents the lower envelope of a family of acceptance wage curves which differ because workers vary in their attitudes regarding risk.

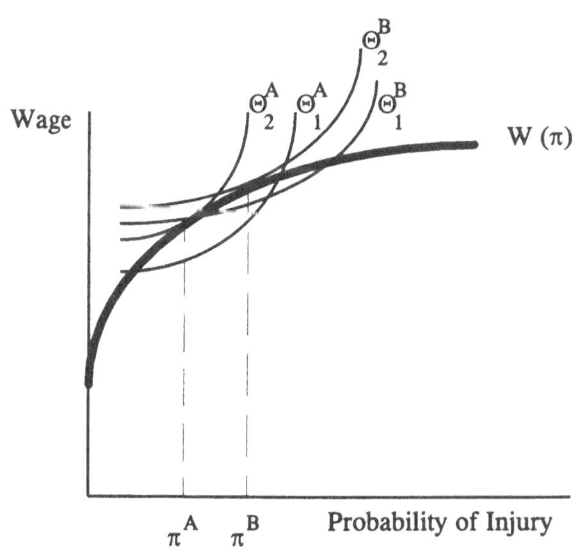

Figure 2.1 Worker Equilibrium

Firms

Whereas the individual employee must decide on the level of injury risk to accept the employer must decide on the level of injury risk to offer. Safety programs are costly. An employer must anticipate corresponding benefits, such as greater output,[2] lower pay for workers, smaller disability insurance premiums, or lower fines for violating government safety standards. The economic problem confronting the employer, then, is to choose the combination of capital, labor, and workplace safety that maximizes expected profit subject to $\pi \geq 0$. To reduce the complexity of the discussion we assume a standard safety device exists that monotonically decreases work-related injuries. We refer to the generic device, which represents all the firm's safety enhancing measures, as safety equipment. Expected revenue is a function of the product price, the probability of a work-related injury, the number of units of capital, and the number of workers. Total expected input cost includes the expected wage bill, the cost of safety equipment, the cost of capital, the cost of workers' compensation insurance, the expected fines for (possibly) violating government workplace safety standards, and any fixed costs of production.

Algebraically, expected profit is

$$\begin{aligned}\bar{g} = {} & R(p_o, S(\pi), n, k; \mu) - W(\pi)(1-\pi)n - p_s S(\pi) \\ & - p_k k - P_b(\pi) bn - V(\pi, e) - c,\end{aligned} \tag{2.3}$$

where

$\bar{g}$	$\equiv$	expected profit,
$R(\cdot)$	$\equiv$	the expected revenue function,
p_o	$\equiv$	the per-unit price of output,
s	$\equiv$	the quantity of safety equipment with $s = S(\pi)$; $S(\cdot)$ is the safety production function and $S' < 0$,
n	$\equiv$	the number of workers,
k	$\equiv$	the quantity of capital,

μ	$\equiv$	a parameter representing the efficiency of safety equipment in the production of output $(\partial^2 R/\partial s \partial \mu > 0)$,
p_s	$\equiv$	the per-unit price of safety equipment,
p_k	$\equiv$	the per-unit price of capital,
$P_b(\pi)$	$\equiv$	the per-unit (of benefits) price of workers' compensation insurance, $P_b' > 0$,
$V(\pi, e)$	$\equiv$	the expected fine for violating safety and health standards with $\partial V/\partial \pi > 0$, $\partial V/\partial e \geq 0$, and $\partial^2 V/\partial \pi \partial e > 0$,
e	$\equiv$	the level of OSHA enforcement activities, and
c	$\equiv$	total fixed costs.

Product price increases revenue at a constant rate while safety equipment, capital, and labor each increase revenue at decreasing rates. All cross derivatives among safety equipment, capital, and labor are positive.

Managers influence expected profit through their decisions on hiring labor, purchasing capital, and reducing injury risk (purchasing safety equipment). All three decisions must be made jointly to maximize expected profit; decreasing risk increases the productivity of labor and capital, while increasing labor and capital increases the net benefits from injury risk reduction. By differentiating equation (2.3) with respect to n, k, and π, setting each result equal to zero and rearranging terms, we can show the optimal usage of each input occurs when

$$\frac{\partial R}{\partial n} = W(\pi)(1-\pi) + P_b(\pi)b, \tag{2.4}$$

$$\frac{\partial R}{\partial k} = p_k, \text{ and} \tag{2.5}$$

$$\frac{\partial R}{\partial S}\frac{\partial S}{\partial \pi} - \left(\frac{\partial W}{\partial \pi}(1-\pi) - w\right)n - \frac{\partial P_b}{\partial \pi}bn - \frac{\partial V}{\partial \pi} = p_s \frac{\partial S}{\partial \pi}. \tag{2.6}$$

Firms increase their use of labor and capital until the expected marginal revenue product of each input equals its expected marginal cost. In addition, firms reduce workplace hazards until the marginal benefits of safety — greater output, lower wages,[3] lower insurance costs, and smaller government fines for workplace hazards — equal the marginal costs of greater safety equipment. Because the output effect of safety equipment varies among workplaces the marginal benefits of reducing work-related health hazards differ among firms, in turn causing the optimum level of safety to vary. Firms where safety equipment is highly productive reduce hazards more than firms where safety equipment is less productive.

Similar to the situation for workers, firms' decisions regarding safety can be viewed graphically. A firm's offer wage function (isoprofit curve) shows the tradeoff between wages and risk at a constant level of expected profit with capital and labor used in optimal quantities. To keep the same level of expected profit wages must fall as safety increases to compensate for the added cost of purchasing safety equipment; offer wage functions slope upwards. Firms with greater costs of producing safety require a greater wage reduction to lower workplace hazards than firms with smaller costs of producing safety, all else equal; the firm with the higher marginal cost of producing safety will have a more steeply sloped offer wage function at a given wage and level of risk than the firm with the lower marginal cost. Finally, profits rise as wages fall implying the lower the offer wage function the higher the expected level of profit.

Figure 2.2 shows the market wage function and offer wage functions for two companies, Clarke Brothers Computer and Darco Explosives. As can be seen, Clarke Brothers Computer maximizes expected profit by offering workers a level of safety equal to $(1-\pi^C)$, the level where the offer wage function is just tangent to the hedonic wage function. Because its costs of improving safety are greater, Darco Explosives maximizes expected profit by offering less safety, $(1-\pi^D)$, but paying higher wages than Clarke Bothers to compensate workers for the greater chance of injury. With a sufficiently large number of diverse firms each point on the hedonic wage function represents a point of tangency for some company or companies. The hedonic wage

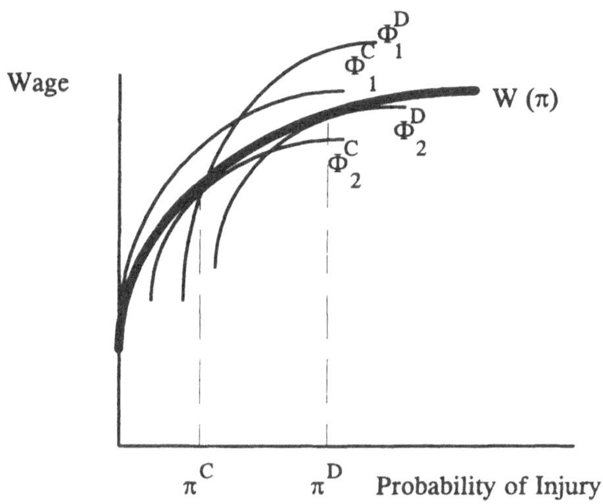

Figure 2.2 Firm Equilibrium

function represents an upper envelope of a family of offer wage curves that differ because of the variation in the ability of firms to produce safe work environments.

Labor Market Equilibrium

Workers sort into a given job type (injury risk) based on the market (hedonic) wage function and their aversion to risk. Firms supply a given type of workplace based on the market wage function and their ability to produce a safe work environment. The hedonic wage function equilibrates the supply and demand for labor along the entire injury risk spectrum. If the supply of labor at a given risk is greater than the demand for labor wages fall. The wage decrease entices some firms to expand their safety efforts to take advantage of the lower wage and induces some workers to accept higher risk to make up for their reduction in income, thereby reducing the excess supply of labor. The movements of firms and workers along the risk spectrum are the market forces determining wages at all levels of safety.

The slope of the acceptance wage function measures the wage a worker is willing to sacrifice to expand safety a small amount and, therefore, provides a dollar figure of worker's willingness to pay implicitly for safety improvements. At the same time the slope of the isoprofit curve measures the reduction in wages required by a firm to compensate for the higher costs of installing additional safety equipment. As can be seen in Figure 2.3, the hedonic wage function maps out a set of tangencies between workers' acceptance wage functions and firms' isoprofit curves or offer wage functions.

Although firms could expand safety programs to reduce workplace hazards below the various levels shown in Figure 2.3 the benefits would be less than the costs. The small wage reduction would not compensate them for their added expenses. Workers could likewise reduce their likelihood of injury by simply choosing to work in safer establishments. They choose not to, however, because the wage sacrifice exceeds the value they place on greater safety. This is not to say workers dislike greater safety. They simply like *both* safety and income and willingly make tradeoffs between them. In equilibrium, the monetary sacrifice workers are willing to make for additional safety just equals firms' costs of providing additional safety.

Figure 2.3 Hedonic Labor Market Equilibrium

WORKPLACE SAFETY POLICY

Workplace safety in the United States is crucially tied to two public policy programs: (1) states' workers' compensation insurance and (2) the federal government's occupational safety and health standards. Passage of WC laws began in the United States near the turn of the century and continued until 1949 when all states had enacted programs. Workers' compensation insurance laws drastically altered the liability standards for industrial safety. Before WC a worker had to prove employer negligence to collect damages for an injury. Further, an employer could avoid paying both medical and income benefits by demonstrating in court that the worker's injury was the worker's fault, was caused by an ordinary hazard of employment, or was caused by a coworker's carelessness. Passage of WC made firms strictly liable for industrial injury. Regardless of fault employers or their insurance carriers must pay an injured worker's medical bills and a portion of his or her lost income while out of work. Proponents of WC believed the change in liability would shift accident costs to firms and encourage greater safety efforts.

For almost sixty years, the WC system appeared to be increasing industrial safety; occupational injury and death rates steadily declined. With the gradual increase in injuries in the mid 1960s the period of relative contentment with just requiring insurance coverage for injured workers ended. A push to establish a set of federal health and safety standards began, which culminated in passage of the Occupational Safety and Health Act of 1970. The Act placed the power to create and enforce safety and health standards in a separate agency of the department of labor, the Occupational Safety and Health Administration. Many believed that by establishing minimum standards for health and safety efforts backed by fines for noncompliance OSHA would dramatically improve industrial safety. The most favorable forecast predicted OSHA would cut workplace injuries in half.[4]

A Theoretical Examination of OSHA

Most OSHA standards require firms install specific types of safety equipment or institute specific safety programs. Besides specification standards OSHA also uses performance standards to set maximum levels of exposure to particular hazards. Both specification and performance standards impose considerable financial burdens on firms. The National Association of Manufacturers calculated OSHA standards cost about $103,000 (1993 dollars) for an average firm with 1–100 employees and $1,026,000 for an average firm with 501–1,000 employees (Smith 1976). Within certain industries the costs are even larger. The 1978 cotton dust standard, for instance, increased operating expenses in the textile mill products industry by more than $50 million per year (Viscusi 1992). Clearly, only if regulations are backed by financial penalties for noncompliance will firms install the safety equipment or institute the safety programs dictated by OSHA.

On a conceptual basis the easiest way to view OSHA's impact on injuries is through its impact on expected profits (equation 2.3). Firms compare the costs of complying with health and safety standards to the costs of not complying. By inspecting more firms or levying higher fines for a given class of violations OSHA increases the costs of noncompliance. The expansion in enforcement raises expected fines for workplace health and safety hazards $(\partial^2 V/\partial\pi\partial e > 0)$ increasing the marginal benefit of safety (the left-hand side of equation 2.6). All else equal, higher penalties encourage firms to eliminate workplace hazards. The direct impact of OSHA is via simply eliminating workplace hazards. However, OSHA also influences both the hedonic wage function and the price of output. Wages must adjust to encourage workers to match the movement of firms to higher safety levels and the higher costs of safety equipment must raise product prices.

Figure 2.4 illustrates the direct impact of OSHA enforcement on workers and firms. For purpose of discussion we assume more inspections and higher fines for noncompliance eliminate extreme workplace hazards causing the highest level of injury risk to fall from π^{max} to π^{dir}.

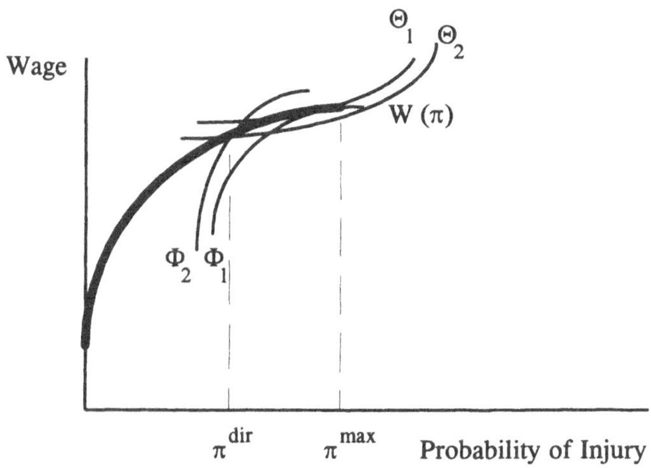

Figure 2.4 The Direct Impact of OSHA on Workplace Safety and Economic Welfare

The offer wage function through π^{dir} represents lower profits than the offer wage function tangent to π^{max} $(\Phi_1 > \Phi_2)$. The expansion in safety via OSHA's direct effect reduces firm profitability — the wage decrease does not compensate for the increase in safety expenses.

Similar to firms, the asking wage function through π^{dir} represents lower worker welfare than the asking wage function just tangent to π^{max} $(\Theta_1 > \Theta_2)$. The decrease in injury risk also reduces worker welfare; workers value the small improvement in safety less than the large drop in wages.

Ultimately the movement of workers and firms to π^{dir} alters the relationship between wages and risk. By installing additional safety equipment firms raise worker productivity $(\partial^2 R/\partial n \partial s > 0)$, lower wages $(\partial W/\partial s < 0)$, and reduce WC insurance premiums $(\partial P_b/\partial s < 0)$. With productivity rising and per-unit labor costs falling firms attempt to hire more workers. Higher production costs force only a few firms out of business, and the increased competition for labor raises wages throughout the risk spectrum. In most cases, to restore equilibrium the hedonic wage function also flattens. The drop in risk compensation encourages workers to match the movement of firms toward greater safety, but also lowers firms' incentives to expand safety.

As seen in Figure 2.5 OSHA enforcement activities raise wages, lower risk compensation, and eliminate high risk jobs. At a minimum, changes in the relationship between wages and risk alter the conclusion that safety and health standards necessarily reduce worker welfare. The rise in the wage function may counteract the wage drop as workers move to safer employment.[5] The new hedonic wage function also points to the need to view OSHA in a complete equilibrium context. The decrease in risk compensation reduces while higher expected fines raise the economic incentive for firms to improve safety. As we show in Chapter 4 because of the change in the hedonic wage function OSHA exposes more workers to moderate workplace health and safety hazards and eliminates fewer extreme workplace dangers, or $\pi^{OSHA} > \pi^{dir}$. The resorting of workers and firms along the risk spectrum mitigates to a substantial extent OSHA's initial effectiveness in reducing work-related accidents and diseases.

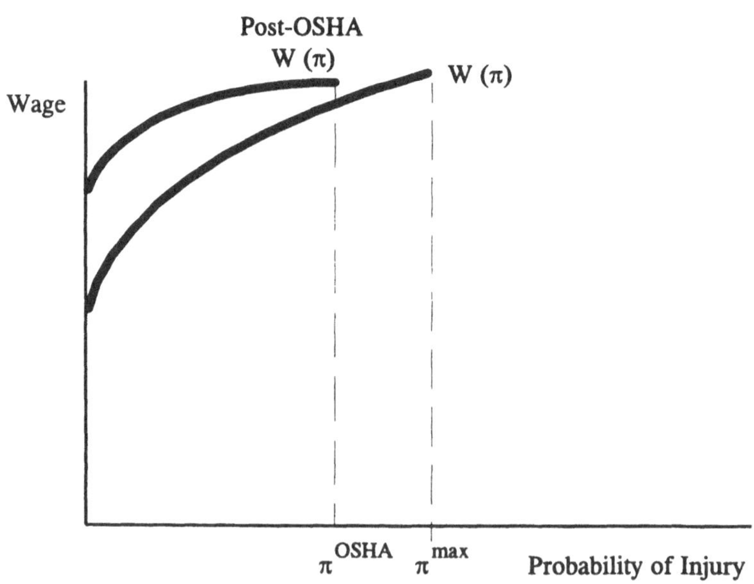

Figure 2.5 The Impact of OSHA on the Hedonic Wage Function

Econometric Estimates of OSHA's Impact

From the theoretical examination of OSHA, safety and health standards should reduce the frequency and the severity of injury. The conditions necessary for changes in the hedonic wage function to produce a perverse outcome appear remote (Rea 1981 and Viscusi 1979). Table 2–1 summarizes the major empirical studies examining OSHA's impact on injuries. The studies vary widely in terms of samples used, risk measurements investigated, and OSHA enforcement activities examined. In broad terms economists have explored three aspects of OSHA: (1) program effects: safety improvements caused by introducing national safety and health standards, (2) deterrence effects: safety improvements caused by firms facing higher probabilities of inspection or higher fines for violations, and (3) abatement effects: safety improvements caused by OSHA inspections.

Program Effects. The most straightforward method to determine the impact on safety from introducing OSHA is to compare injury rates before and after passage of the Act in 1970. Unfortunately for researchers the change in safety policy was accompanied by a change in the method used by the BLS to monitor injuries, which makes a direct pre- versus post-OSHA comparison using national injury rate statistics invalid.

Two studies have tried to overcome the injury measurement problem by calculating the frequency and severity of injuries using information on WC claims. Based on California data from 1948 to 1975, Mendeloff (1979) found that OSHA reduced "caught in or between" injuries and falls and slips; had no impact on "struck by" injuries or all disabling injuries; and *raised* the number of strains. Based on New York data from 1964 to 1976, Curington (1986) found that overall OSHA reduced only "struck by machine" injuries. On a more disaggregated basis, however, Curington found OSHA lowered the severity of injuries in five industries, the frequency of "struck by machine" injuries in nine industries, and the severity of "caught in machine" injuries in six industries. He also discovered OSHA *raised* the total frequency of injury in five industries (only one industry experienced a reduction in total injuries), raised the severity of injury in three indus-

Table 2-1. The Impact of OSHA on Injuries

Study	*Sample*	*Risk Measurement*	*OSHA Enforcement Measurement*	*Impact*
Mendeloff (1979)	Time-series of manufacturing injury rates from 1948 to 1975	California Department of Industrial Relations Frequency of:	Predicted injuries based on pre-OSHA data versus actual injuries	
		(1) Total disabling injuries		(1) No impact
		(2) Caught in or between injuries		(2) Decrease 1973-1975
		(3) Struck by injuries		(3) No impact
		(4) Strains		(4) Increase 1972-1975
		(5) Falls and slips		(5) Decrease 1974 and 1975
Smith (1979)	Cross-section of plants from 1972 to 1974	Bureau of Labor Statistics (BLS) Frequency of lost workday injuries	OSHA inspections (1=inspected early in the year, 0=inspected late in the year)	Plants inspected in 1973: Injuries fell 7.1%[a] in 1973 and 11.6%[a] in 1974 Plants inspected in 1974: No significant reduction
Viscusi (1979)	Time-series/cross-section of 2-digit industries from 1972 to 1975	BLS Log odds of injuries and illnesses per worker	Inspections per 100,000 workers in years t, t-1, t-2, and t-3	Individually and jointly insignificant
			Proposed penalties per 1,000 workers in years t, t-1, t-2, and t-3	Individually and jointly insignificant
Cooke and Gautschi (1981)	Time-series/cross-section of Maine manufacturing plants from 1970 to 1976	Maine Industrial Accident Commission Change in average days lost from injury	OSHA citations per plant	OSHA citations reduced average days lost by 23.0%[a] in plants with 200 or more workers and 37.0%[a] in plants with 300 or more workers

McCaffrey (1983)	Cross-section of plants from 1976 to 1978	BLS Frequency of lost workday injuries	OSHA inspections (1=inspected early in the year, 0=inspected late in the year)	No significant reductions in 1976, 1977, and 1978
Bartel and Thomas (1985)	Time-series/cross-section of 3-digit industries in 22 states from 1974 to 1978	BLS Log lost workdays per 100 workers	Penalties per inspection/ employees per firm	No significant impact on injuries
Curington (1986)	Time-series/cross-section of New York manufacturing industries from 1964 to 1976	New York Workmen's Compensation Board Frequency of WC claims and days of impairment: (1) all injuries, (2) caught in machine injuries, and (3) struck by machine injuries	Passage of OSHA (1=post-OSHA, 0=pre-OSHA)	Pooled sample: OSHA reduced struck by machine injuries 13.1%[a] By industry: OSHA decreased the severity of injury in 5 industries, the frequency of struck by machine injuries in 9 industries, and the severity of caught in machine injuries in 6 industries
Viscusi (1986a)	Time-series/cross-section of 2-digit manufacturing industries from 1973 to 1983	BLS (1) Frequency of injuries and illnesses (2) Frequency of lost workday injuries and illnesses (3) Rate of lost workdays	Inspections per production worker in year t and t-1	Total percentage impact: (1) -0.04[b] (combined) -2.6[a] (year t-1) (2) -1.5[b] (combined) -3.6[a] (year t-1) (3) -4.7[b] (combined) -6.1[a] (year t-1)
			Assessed penalties per production worker in year t and t-1	Insignificant across all three risk measures
Scholz and Gray (1990)	Time-series/cross-section of manufacturing plants from 1979 to 1985	BLS (1) Percentage change in lost workday injuries	Change in the expected probability of an inspection with penalty years t, t-1, and t-2	A 10% increase in inspections reduces injuries by 1.61% and lost workdays by 0.93%

Table 2-1. *(Continued)*

Study	*Sample*	*Risk Measurement*	*OSHA Enforcement Measurement*	*Impact*
Scholz and Gray (1990), *cont.*		(2) Percentage change in lost workdays	Change in the expected penalty given an inspection with penalty in years t, t-1, and t-2	A 10% increase in average penalties reduces injuries by 0.88% and lost workdays by 0.50%
			Inspection with penalty in years t, t-1, t-2, and t-3 (1=yes, 0 otherwise)	Note: Most OSHA enforcement coefficients are negative and significant
Ruser and Smith (1991)	Time-series/cross-section of plants from 1979 to 1985	BLS Frequency of lost workday injuries	OSHA inspections (1=inspected early in the year, 0=inspected late in the year)	Insignificant in year t and t+1
			Frequency of inspections per establishment by state, 2-digit industry, and 9 establishment size classes in year t and t-1	Insignificant both individually and jointly
Gray and Scholz (1993)	Time-series/cross-section of manufacturing plants from 1979 to 1985	BLS (1) Percentage change in lost workday injuries (2) Percentage change in lost workdays	Inspection with penalty years t, t-1, t-2, t-3	An inspection with penalty reduces injuries by 22% and lost workdays by 20% over the following 3 years Note: Most inspection variables are negative and significant

[a]Significant at the 5 percent level, 1-tail test
[b]Significance levels not specified

tries, and raised the frequency of "caught in machine" injuries in two industries. The Mendeloff and Curington studies provide a mixed picture of OSHA's initial effectiveness in encouraging safer workplaces. Neither Mendeloff nor Curington found that OSHA significantly improved total workplace safety, and both found some injuries rose after the start of OSHA.

No one seriously believes OSHA's initial activities were sufficiently severe to reduce worker safety efforts enough to counteract firms' efforts to improve safety, implying the estimated growth of injuries may be spurious. Because both Mendeloff and Curington used a limited time period after OSHA's passage to determine its impact on injuries their estimates may be fragile. Mendeloff, for instance, found some of his results showing a positive impact of OSHA on workplace safety not to be robust when updating his study with additional post-OSHA data (as reported in Viscusi 1986a).

Relying on WC claims to determine injury frequency and severity may produce even more serious difficulties with the Mendeloff and Curington studies. As we discuss more fully in Chapter 5, increases in WC benefits cause injury frequencies calculated using WC claims to rise even if actual safety improves. The economic incentives for workers to apply for benefits as WC becomes more generous causes reported injuries (claims) to diverge substantially from true injuries. Curington tried to control for WC by including a measure of income replacement in his empirical model, but the measure may not have adequately captured the complete relationship between reported injuries and benefits or the complete range of changes in New York's WC program from 1964 to 1976. Mendeloff made no attempt to control for WC benefit generosity.

Deterrence Effects. Other studies in Table 2-1 have tried to find OSHA's impact on injuries by including an enforcement measurement that varies over establishments, industries, or time. Viscusi (1979, 1986a) and Bartel and Thomas (1985) used industry data while Ruser and Smith (1991) and Scholz and Gray (1990) used plant data to determine the deterrence effects from inspections and penalties. All else equal, workplace injuries should fall as the frequency of inspections or the average fine per inspection rises. Expanding the likelihood of detection and financial penalty should deter firms from violating health and safety standards.

The two studies by Viscusi cited in Table 2-1 have used the identical empirical method to determine deterrence effects of OSHA inspections and penalties, but have come to different conclusions. The first found no significant impact from OSHA enforcement activities while the second discovered the frequency of inspections reduced both injuries and lost workdays. Because the results showed no evidence of increasing OSHA effectiveness over time, Viscusi credited the two studies' varying levels of statistical significance to the differing sample sizes. With more years to examine the later Viscusi study was able to detect a statistically significant, but fairly small, impact from OSHA enforcement efforts. At most, from 1973 to 1983 OSHA inspections lowered the frequency of lost workday injuries by 1.5-3.6 percent.

The most puzzling aspect of Viscusi (1986a) is his finding that current OSHA inspections raised while lagged inspections lowered the frequency and severity of injuries. One might suspect econometric problems with simultaneity bias; if OSHA targets industries with deteriorating workplace safety then the positive relationship between current inspections and current injuries might reflect the impact of injuries on OSHA activities. Nevertheless, specification tests found no simultaneity, leaving the impact of current inspections on injuries a mystery. The sensitivity of the signs of the inspection coefficients to changing samples is also disturbing. Although statistically insignificant, both current and lagged inspections reduced the frequency and severity of injuries 1973-1979 but raised the frequency and severity of injuries during 1980-1983.

Unlike Viscusi, Bartel and Thomas did not find OSHA activities significantly reduced workplace hazards. Enforcement efforts expanded industry compliance with health and safety standards but greater compliance failed to lower injuries. Bartel and Thomas interpreted their results as indicating the general ineffectiveness of OSHA's approach of mandating the physical characteristics of acceptable plant and equipment. OSHA can force firms to abide by their dictates but OSHA's dictates apparently had little to do with improving safety during 1974-1978.

The two studies in Table 2-1 using establishment data likewise came to quite different conclusions regarding OSHA's deterrence effects. Ruser and Smith found no evidence that more frequent inspections improved workplace safety. Scholz and Gray found that inspections and penalties dramatically reduced injuries; their results revealed that a 10 percent increase in OSHA inspections lowered lost workday accidents and diseases by 1.61 percent while a 10 percent rise in average penalties reduced injuries by 0.93 percent. In total, the Scholz and Gray estimates imply OSHA's enforcement activities have lowered workplace injuries by 10 to 16 percent, which is high relative to previous estimates. Scholz and Gray argued that their results reflect the less than representative nature of their sample. The firms Scholz and Gray studied are more hazardous and more frequently inspected than most manufacturing firms and so are more responsive to OSHA enforcement efforts. Using the same empirical framework as Viscusi (1986a) Scholz and Gray found a 10 percent increase in OSHA inspections lowered injuries on average by 1.27 percent in their sample. In contrast, Viscusi found only a 0.16 percent reduction in lost workday injuries for all of manufacturing.

Abatement Effects. The remaining studies in Table 2-1 have examined the abatement impact of OSHA enforcement activities on workplace safety. Smith (1979), McCaffrey (1983), and Ruser and Smith (1991) determined the effectiveness of OSHA inspections by comparing lost workday injuries in plants inspected early in the year to lost workday injuries in plants inspected late in the year. Firms inspected first have more time to eliminate workplace hazards and, therefore, should have fewer injuries if inspectors correctly identify health and safety hazards. By examining only inspected companies, the three studies indirectly controlled for factors causing an inspection, such as deteriorating workplace safety, that might confound estimates of OSHA's effectiveness.

In general, the studies did not find inspections significantly lowered lost workday injuries. Although Smith discovered inspections in 1973 reduced injuries, he did not detect a significant OSHA abatement effect in 1974. Likewise, McCaffrey found no impact on lost workday injuries from 1976 to 1978, and Ruser and Smith found no impact from

1979 to 1985. As Viscusi (1986a) speculated, the 1973 results may have either reflected an initial overreporting of injuries caused by the start of OSHA or a regression-to-the-mean effect. In either case the estimated positive impact on workplace safety in 1973 is spurious.

The studies by Smith, McCaffrey, and Ruser and Smith may have underestimated OSHA's effectiveness because they lumped together plants violating with plants not violating health and safety standards. On average, injuries should fall only in firms as they move into compliance with OSHA. By not differentiating between inspections resulting in fines and inspections not resulting in fines the three studies may have understated OSHA's power to improve safety in particularly hazardous workplaces.

Two studies have specifically examined the impact of violations on plant safety; both found OSHA significantly reduced hazards. Cooke and Gautshi (1981) examined a sample of Maine manufacturing plants during 1970–1976 and discovered OSHA citations lowered lost workdays by 24 percent in plants with more than 200 workers and lowered lost workdays by 37 percent in plants with more than 300 workers. Gray and Scholz (1993) found an equally impressive impact of OSHA using a large national cross-section of establishments from 1979 to 1985. Gray and Scholz found that an OSHA inspection with penalty decreased lost workday injuries by 22 percent over three years.

In total, the econometric evidence summarized in Table 2–1 suggests a positive but modest impact of OSHA enforcement activities on workplace safety. Although inspections resulting in penalties produced marked improvements in safety, only a third of all inspections result in fines and few firms are ever inspected. With a 10 percent inspection rate, OSHA abatement efforts reduced total injuries by less than 1 percent per year. Injuries may have been further reduced because the threat of inspections and fines encouraged firms not yet inspected to install additional safety equipment or institute new safety programs. Based on Viscusi's (1986a) work the deterrence effect of OSHA lowered injuries by another 1.5–3.6 percent. At most, then, the econometric evidence shows OSHA having reduced injuries by 5 percent.[6]

A Theoretical Examination of WC

A major goal of WC is to provide firms with incentives to invest optimally in health and safety programs (Berkowitz and Burton 1970). By shifting accident costs from workers to firms WC tries to encourage firms to spend more on safety enhancing measures. As can be seen in equation 2.6 firms consider the savings from lower WC premiums when deciding on the optimal level of safety to provide. The term $[(\partial P_b/\partial\pi)bn]$ reflects the reduction in WC premiums from higher levels of safety. Notice the size of the reduction is directly related to the responsiveness of insurance prices (P_b) to changes in workplace hazards (π), the size of the income benefits mandated by state law (b), and the level of employment (n). All else equal, if insurance prices fall as safety rises higher WC benefits expand the marginal benefits from safety, and firms offer safer workplaces.[7] We call the effect via insurance prices the direct impact of WC because it assumes benefits influence neither the hedonic wage function nor the revenue function. Clearly, both firms' revenues and the hedonic function must change to restore equilibrium to the labor market. At the very least wages must adjust to encourage workers to match the movement of firms to higher safety levels, and prices must rise to reflect additional labor costs.

Wages also change because WC benefits alter workers' decisions regarding safety. When choosing the level of employment risk to accept workers consider both the utility and income loss from injury (See equation 2.2). By reducing the economic loss from an accident or a disease higher WC benefits reduce the marginal cost from injury and, thereby, encourage workers to accept less safe employment if wages remain constant. The size of the reduction in the marginal cost of injuries depends on the worker's marginal utility of income if injured. The higher the initial level of benefits or alternative sources of income the smaller the decline in the marginal costs of injury and the smaller the increase in risk. Likewise, the greater the health consequences from injury the smaller the decline and the smaller the increase in risk. Changes in the hedonic wage function must reconcile the disparate interests of workers and firms.

On a theoretical basis one cannot determine whether higher WC benefits expand or contract workplace safety. Workers choose less safety at the same time firms choose more safety. If the impact of higher WC benefits on workers is sufficiently great then changes in the hedonic wage function counteract firm incentives to provide additional safety, and the frequency of injury rises. On the other hand, if the impact of higher benefits on firms is sufficiently great then changes in the hedonic wage function counteract worker incentives to choose higher risk jobs, and the frequency of injury falls. Safety is more likely to improve the more closely WC insurance premiums reflect workplace hazards, the higher the initial level of benefits, the greater nonlabor income (including alternative sources of disability income such as SSDI), and the larger the health consequences from a work-related injury.

Because the impact of WC benefits on injuries is uncertain the impact on the hedonic wage function is also uncertain. In general, as shown in Figure 2.6, higher benefits reduce wages at every level of safety. A rise in nonwage compensation increases labor supply while decreasing labor demand, thereby lowering wages. As we demonstrate later, if insurance prices at least partially reflect workplace hazards across firms, then more generous WC benefits will reduce injuries by eliminating the most extreme workplace health and safety hazards and by relocating workers and firms toward greater safety. Although higher WC benefits reduce the economic costs of injuries workers sort toward greater safety because the compensation for risk falls. In short, an increasingly generous WC insurance program raises workplace safety by lowering, flattening, and truncating the hedonic wage function.

Econometric Estimates of WC's Impact

Table 2-2 summarizes the major studies exploring WC's impact on workplace safety. Studies of workers' compensation insurance effects on the workplace vary tremendously in terms of samples used, risk measurements investigated, and characteristics examined. In general terms they address three questions: (1) Did passage of state WC laws

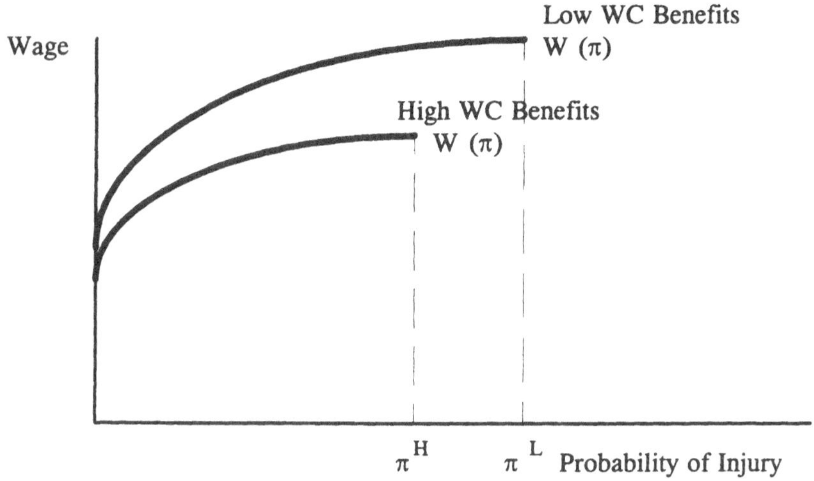

Figure 2.6 The Impact of WC Benefits on the Hedonic Wage Function

improve workplace safety? (2) Do higher income benefits reduce the frequency or the severity of injuries? and (3) Does experience rating encourage firms to reduce workplace hazards?

Introduction of WC. In many respects, determining the impact from changing the liability system for workplace injuries should be more straightforward than determining the safety impact from the introduction of OSHA. The introduction of WC did not change firms' requirements for reporting injuries so one can compare a state's injury rate before and after WC to determine its influence on safety. One can also examine the variation in injuries across states to identify the impact of WC because until 1948 not all states had enacted WC legislation. Nevertheless, measuring the safety effectiveness of WC is still difficult. No consistent measure of workplace injuries exists over the almost fifty years from the first to the last state's enactment of WC, and limited information is available on intervening factors.

Two studies have examined the safety effects from introducing WC. Chelius (1976) measured the impact of WC on safety using data on machinery deaths per worker from 1900 to 1940. Although not all

Table 2-2. The Impact of Workers' Compensation on Injuries

Study	*Sample*	*Risk Measurement*	*Workers' Compensation Benefit Variable*	*Impact*
Chelius (1974)	Cross-section of establishments from 13 states	Bureau of Labor Statistics (BLS) Frequency of disabling injuries	Instrumental variable representing an index of WC benefits by state and 2-digit industry	Positive[a]
			Length of waiting period	Insignificant
Chelius (1976)	Time-series/cross-section of 26 states from 1900 to 1940	U.S. Bureau of the Census Machinery death rate by year relative to national average by year	Workers' compensation law (1=law, 0 otherwise)	Negative[a]
Chelius (1982)	Time-series/cross-section of 2-digit manufacturing industries in 36 states from 1972 to 1975	OSHA (1) Lost workdays (2) Frequency of lost workday cases (3) Days lost per case	(B) Income replacement rate (W) Length of waiting period	Elasticities 1 / 2 / 3 B: 0.08 / 0.14[a] / 0.09[a] W: -0.02 / -0.03[a] / 0.01
Worrall and Appel (1982)	Time-series of Texas WC claims from 1958 to 1977	National Council on Compensation Insurance (NCCI) (1) Temporary total indemnity claims to medical only claims (2) All indemnity claims to medical only claims (3) Cost of temporary total indemnity claims to the cost of medical only claims	Injured workers' after tax wage replacement for temporary total disabilities	Elasticities: (1) 0.614[a] (2) 0.465[a] (3) 0.990[a]

		(4) Cost of all indemnity claims to the cost of medical only claims		(4) 0.960^a
Butler (1983)	Time-series/cross-section of 15 industries in South Carolina from 1940 to 1971	South Carolina Industrial Commission Frequency of: (1) fatalities (2) permanent dismemberment or disfigurement injuries (3) permanent partial injuries (4) temporary total injuries (5) an injury index	Log of average annual benefits for: (a) fatalities (b) permanent dismemberment or disfigurement injuries (c) permanent partial injuries (d) temporary total injuries Log of: (e) a benefit index (f) expected weekly benefits for the average worker	Estimates vary by dependent variable and benefit measurement Elasticities for Aggregated WC Measures e / f 1: 1.13 / 5.20^a 2: 0.36^a / na 3: 0.33^a / na 4: 0.13 / na 5: 0.29 / 1.02
Butler and Worrall (1983)	Time-series/cross-section of 35 states from 1972 to 1978	NCCI Log of claim frequency for: (1) temporary total (2) minor permanent partial (3) major permanent partial	Log of average weekly benefits for: (a) temporary total (b) minor permanent partial (c) major permanent total (d) length of waiting period	Elasticities 1 / 2 / 3 a: 0.37 / 0.60^a / 1.71^a b: 0.02 / 0.26^a / -0.19 c: 0.32 / -0.45^a / 0.64 d: -0.33^a / -0.17^a / -0.02
Chelius (1983)	Time-series/cross-section of 2-digit manufacturing industries in 28 states from 1972 to 1978	OSHA (1) Lost workdays (2) Frequency of lost workday cases (3) Days lost per case	(B) Income replace rate for temporary total disabilities (W) Length of the waiting period	Elasticities 1 / 2 / 3 B: 0.16^a / 0.22^a / -0.05 W: -0.02 / -0.04 / 0.01

Table 2-2. *(Continued)*

Study	*Sample*	*Risk Measurement*	*Workers' Compensation Benefit Variable*	*Impact*
Chelius and Smith (1983)	15 2-digit manufacturing industries across 37 states in 1979	BLS The difference across large and small establishments in the frequency of lost work-day cases	Temporary total benefits for the average worker	Across the 60 industry and size classes 32 estimates are negative, 5 significant
Leigh (1985)	Time series/cross section of individuals from 1977 to 1980	Panel Study of Income Dynamics (PSID) Receipt of WC benefits (1 = receipt, 0 otherwise)	Potential temporary total weekly benefits	0.3 (sample mean elasticity controlling for time and state effects)
Worrall and Butler (1985); Butler and Worrall (1985)	Claims filed by men for low back injuries in Illinois	NCCI Duration of temporary total disabilities	Log of the income replacement rate for temporary total disabilities Log of temporary total benefits	Duration elasticities: 0.187^a, 0.374^a (correcting for duration dependence) 0.183 (correcting for unobserved heterogeneity)
Ruser (1985)	Time-series/cross-section of 25 3-digit industries across a maximum of 41 states from 1972 to 1979	BLS (1) Frequency of injuries (2) Frequency of lost work-day injuries, excluding fatalities	(B) Average weekly benefits for temporary total disabilities (P) Average proportion of compensable lost workdays	Elasticities 1 / 2 B: 0.275^a / 0.313^a P: -0.365 / 0.509
Fishback (1987)	Time-series/cross-section of 23 leading coal states from 1903 to 1930	U.S. Bureau of Mines Frequency of coal mining fatalities	Employer liability laws (1 = laws present and no WC, 0 otherwise) Exclusive state WC insurance fund (1 = yes, 0 otherwise)	Percentage change: 17.3^a 23.0^a

			Non-exclusive state WC insurance fund (1=yes, 0 otherwise)	20.2[a]
			WC without a state insurance fund (1=yes, 0 otherwise)	15.6
Butler and Worrall (1988)	Establishment data drawn from 11 industries across 38 states from 1980 to 1981	NCCI Indemnity losses per million dollars of payroll	Maximum payment for temporary total benefits	A 10% increase in benefits raises insurance costs by 14.69%[c]
Chelius and Kavanaugh (1988)	Maintenance workers in 2 community colleges from the first quarter of 1978 through the third quarter of 1984	WC Injury Reports (1) Lost-time injuries per worker (2) Severe lost-time injuries per worker (more than 7 days duration) (3) Lost workdays per worker (4) Lost workdays from severe injuries per worker (more than 7 days duration)	Low WC benefits (1=70% wage replacement and a 7 day waiting period, 0=100% wage replacement and no waiting period)	(1) Negative[a] (2) Negative[a] (3) Insignificant (4) Insignificant
Worrall and Butler (1988)	Time-series/cross-section of 15 industries in South Carolina from 1940 to 1971	South Carolina Industrial Commission (1) Permanent partial injuries per employee (2) Temporary total injuries per employee (3) An index of all indemnity claims	Expected benefit for the average worker	Elasticities: (1) 2.457[c] (2) 0.160[c] (3) 1.985[c]

Table 2-2. *(Continued)*

Study	*Sample*	*Risk Measurement*	*Workers' Compensation Benefit Variable*	*Impact*
Moore and Viscusi (1989, 1990)	Cross section of individuals from the 1982 PSID	National Institute of Occupational Safety and Health Frequency of fatalities by 1-digit industry and state	Maximum weekly benefits for temporary total disabilities	Elasticities: -0.22[a]
			Expected weekly benefits for temporary total disabilities	-0.48[a]
Johnson and Ondrich (1990)	1,040 permanently partially impaired workers from Florida, New York, and Wisconsin injured in 1970	Survey Data Duration of absence away from work	Scheduled benefits	Duration elasticities: 1.114[c] (Weibull) 1.161[c] (Weibull with gamma heterogeneity) 1.128[c] (Weibull with two-point heterogeneity)
Krueger (1990a)	Time series/cross section of individuals 1983-1984 and 1984-1985	Current Population Survey Initial receipt of WC benefits (1=receipt, 0 otherwise)	Log expected temporary total benefits	Varies by included regressors and sample composition Elasticities range: 0.454 to 0.741[b] 0.428 to 0.993[b] (men) -0.285 to -0.006 (women)
			Length of waiting period	Varies by included regressors and sample composition: an increase from 3 to 7 days lowers recipiency rate by 38.7%[a] (specification #1 for total sample)
			Retroactive period	Approximately 0

Krueger (1990b)	Claims filed in Minnesota for injuries occurring in 1986	Administrative Records Log duration of temporary total injuries	Log weekly temporary total benefits	Elasticity: 1.67[a]
Krueger and Burton (1990)	Time-series/cross-section of 29 states in 1972, 1975, 1978, and 1983	NCCI Log of adjusted manual rates (WC premiums) per $100 of payroll	Log of weighted weekly benefits for temporary total, permanent partial, permanent total, and fatal injuries	A variety of empirical specifications — none suggest higher benefits increase injuries
			Log of net WC costs per worker	A variety of empirical specifications — none suggest higher benefits increase injuries
Butler and Worrall (1991)	Time-series/cross-section of 33 states from 1954 to 1981	NCCI (1) Log of expected indemnity costs per worker (2) Log of expected medical costs per worker	Log of temporary total benefits for a typical worker	(1) 0.68[a] (claims reporting moral hazard elasticity) (2) -0.36[a] (risk bearing moral hazard elasticity)
Ruser (1991)	Time-series/cross-section of manufacturing establishments from 1979 to 1985	BLS Frequency of lost workday injuries	Average weekly benefits for temporary total disabilities interacted with establishment size	Varies by empirical technique Elasticities for negative binomial: 0.53[a] (1–99 employees) 0.21[a] (100–249) 0.21[a] (250–499) 0.12[a] (500+)

Table 2-2. *(Continued)*

Study	*Sample*	*Risk Measurement*	*Workers' Compensation Benefit Variable*	*Impact*
Ruser (1993)	Time-series/cross-section of manufacturing establishments from 1979 to 1984	BLS (1) Non-lost-workday injuries per establishment (2) Restricted workday injuries per establishment (3) Lost workday injuries per establishment (4) Fatalities per establishment	Average weekly benefits for temporary total disabilities interacted with establishment size Employment categories: (a) 1–99 (b) 100–249 (c) 250–499 (d) 500 and up	Varies by empirical technique Elasticities for Non-Fixed Effects: 1 2 3 4 a 0.16[a] 0.12 0.57[a] -1.31[a] b 0.08 -0.20[a] 0.28[a] -1.57 c 0.02 -0.37[a] 0.26[a] -0.96 d -0.05 0.34[a] 0.17[a] -1.47[a]
Butler (1994)	U.S. time-series from 1947 to 1990	National Safety Council Log frequency of workplace fatalities	Log of WC costs per payroll source	Varies by empirical technique Elasticities range: -1.56 to -0.36[a]
			Log of the employment share weighted average of 1990 industrial death rates	Varies by empirical technique Elasticities range: -0.06 to -0.04[b]
Curington (1994)	Scheduled permanent partial WC cases closed in New York from 1964 to 1983	New York Workers' Compensation Board Log duration: (1) Severe Impairments (2) Minor Impairments	Workers affected by legislative changes in WC benefits (1=yes, 0 otherwise)	Varies by year of change Elasticities for 1968: (1) 1.08[a] (2) 0.13

[a]Significant at the 10 percent level, 2-tail test
[b]Some estimates significant at the 10 percent level, 2-tail test
[c]Significance levels not specified

machinery deaths are work related and not all workplace fatalities are caused by machinery, the machinery death rate represents the best available overall measure of industrial safety. In contrast, Fishback (1987) handled the problem of a consistent injury rate statistic by investigating the impact of WC on fatalities in a single industry, coal mining. Chelius and Fishback came to quite different conclusions. Chelius found WC lowered machinery deaths, while Fishback found WC raised coal mining fatalities. Neither study is without fault. As Chelius admitted, the machinery death rate is an imperfect measure of workplace safety, but the alternatives are likewise imperfect. A more precise but narrowly drawn injury rate statistic, such as fatalities in a single industry, may miss the general influence of WC on safety. From the available evidence, then, the movement towards strict employer liability for worker injuries had an uncertain impact on workplace safety.

The Growth of Benefits. Econometric evidence on nonfatal injuries is far clearer. As shown by Chelius (1974, 1982, and 1983), Butler (1983), Butler and Worrall (1983, 1988), Leigh (1985), Ruser (1985, 1991, and 1993), Chelius and Kavanaugh (1988), Worrall and Butler (1988), and Krueger (1990a), higher WC benefits increased workplace accidents and diseases. The conclusion is robust to sample composition, injury definition, or benefit measurement. Studies we describe in Table 2-2 examined samples composed of states, industries, establishments, and individuals. They investigated all injuries, lost workday injuries, disabling injuries, and injuries by WC classification.[8] The studies in Table 2-2 judged WC generosity either directly by including cash payments for typical disabilities, all disabilities, or temporary total disabilities or judged WC generosity indirectly by calculating an income replacement rate.

Of all the studies in Table 2-2, only Krueger and Burton (1990) did not find an impact of WC benefits on safety, and only Butler and Worrall (1983) and Ruser (1993) found benefits significantly decreased any category of nonfatal injury. Butler and Worrall and Ruser argued their results reflect the substitution among injury categories, not a general reduction in injuries. Few areas in economics show such consistency in results as those in Table 2-2, which despite vast differences in approach, generally indicate higher WC benefits raised the frequency of workplace accidents and diseases.

The impact of higher benefits on workplace fatalities is only slightly less clear cut. Butler (1983) found more generous WC payments raised fatalities, while Moore and Viscusi (1989, 1990), Ruser (1993), and Butler (1994) found more generous payments lowered fatalities. In his 1983 study Butler used 32 years of industry data from South Carolina to examine the relationship between benefits and fatalities. By limiting the investigation to a single state, Butler indirectly controlled for institutional variations among states that may have influenced the reporting of workplace fatalities. Nevertheless, a single state's experience may not mirror the relationship between benefits and deaths found in most states. The other studies have directly determined the typical outcome of benefit generosity on fatalities by examining broad national cross-sections of workers and establishments. Although they have used quite different measures of fatality risk and employed quite different empirical methods, the econometric studies have arrived at the same conclusion — WC dramatically reduces workplace fatalities. Based on Moore and Viscusi (1990) totally eliminating WC would increase fatal accidents and diseases by 48 percent.

How can more generous WC benefits expand injuries overall but lower fatalities? The economic theory discussed previously provides one possible explanation for an inverse movement in overall injuries and fatalities due to WC. Workers consider both the monetary and nonmonetary costs of injury when deciding on the level of workplace risk to accept. Because the nonmonetary costs of fatalities are so large, higher WC death benefits are unlikely to entice workers to accept greater fatal injury risk. On the other hand, the pain and suffering from some types of less severe injuries may be sufficiently small that more generous WC benefits encourage workers to reduce their safety efforts. In either case the initial reaction of firms to the expanding costs of WC premiums is to invest more heavily in safety. With fatal injuries the direct impact on firms may dominate the limited impact on workers causing deaths to fall, while with nonfatal injuries the direct impact on workers may dominate the impact on firms causing injuries to rise.

The economic incentive of workers to report injuries as WC generosity expands provides an alternative, and perhaps more realistic, explanation of the diverging empirical results on injuries and fatalities. Workers are more likely to report accidents and diseases and file claims for disability benefits when benefits are high, so that even if WC had no influence on safety the rising incentive to file a WC claim would cause a positive relationship between benefits and injuries reported to WC authorities. WC reduces true safety, though, only if workers place a low value on pain and suffering, but for most workers the major cost of a work-related accident or disease is the nonmonetary loss, not the reduction in income. Econometric estimates suggest workers would collectively sacrifice about $50,000 to avoid a single lost workday injury and collectively sacrifice $3–$7 million to avoid a single workplace fatality (Viscusi 1992).

Because of the difficulty of disputing or exaggerating serious injuries or fatalities workers or their families have little reason to file false claims for WC benefits. Workplace fatalities should be relatively free from mismeasurement and, therefore, should most accurately mirror real changes in safety. The diverging impact of WC benefits on fatal and nonfatal injuries may then simply reflect the difference between real and reported workplace accidents and diseases. Higher benefits improve workplace safety reducing both fatal and nonfatal injuries, but reported nonfatal injuries rise as some workers file false or exaggerated claims for WC benefits and others file previously unreported claims because it is now financially worthwhile.[9]

As we show in Chapter 5 the expanding incentive to file for WC benefits can easily swamp the true improvement in safety caused by a more generous WC program. Only one study in Table 2–2 directly addresses the question of reported versus actual injuries. Butler and Worrall (1991) separated reporting effects from injury rate effects using insurance data on indemnity (wage loss) payments and medical costs. In the absence of any change in safety or incentive to report injuries the variability of WC indemnity payments across states or over time should reflect only differences in benefit generosity. A one percent increase in benefits should produce a one percent increase in indemnity payments, all else equal. Holding safety constant, if more generous benefits produce a reporting effect then a one percent increase in WC benefits will cause indemnity payments to rise by more than one percent. Based on changes in medical costs Butler and Worrall found that a one percent increase in WC benefits reduced workplace injuries by 0.36 percent (which is known as the risk bearing moral hazard effect). The 0.36 percent reduction in workplace injuries when WC benefits increase one percent should have lowered indemnity payments by 0.36 percent, but indemnity payments actually rose by 0.32 percent, implying an increase in reported injuries of 0.68 percent (which is known as the claims reporting moral hazard effect).

The expanding incentive to report injuries provides a credible explanation for the rise in accidents and diseases and the fall in fatalities as WC becomes more generous. It may also help explain why Butler (1983) found WC increased fatalities, while Moore and Viscusi and Ruser found WC reduced fatalities. Moore and Viscusi measured fatal injury risk using information from a national census of death certificates, and Ruser gauged fatal injury risk using employer reports to the BLS. In contrast, Butler calculated the rate of fatal injuries using data on closed WC claims. To the extent work-related deaths are difficult but not impossible to dispute Butler's measure of fatal injury risk may also suffer from a residual reporting effect. Families may more frequently and forcefully contest the denial of death benefits as benefit generosity expands, causing a positive relationship between successful (closed) WC death claims and WC benefits. Information from death certificates or BLS reports should be free from reporting effects and, therefore, more accurately mirror true changes in safety.

Reporting effects may also explain the empirical results on injury severity. In an effort to reduce the costs of purchasing WC firms should expand their efforts to reduce both the number and the severity of injuries as WC benefits rise. Nevertheless, with only one exception (Chelius 1983) the studies summarized in Table 2–2 show a positive relationship between WC benefits and injury severity. Chelius (1982) found higher levels of WC increased days lost per case, while Worrall and Butler (1985), Butler and Worrall (1985), Johnson and Ondrich (1990), Krueger (1990b), and Curington (1994) found that higher benefits raised claim duration. The results that WC seems to increase injury severity may reflect real changes in safety or they may simply reflect reduced incentives to return to work after an injury. By reducing the wage loss higher WC benefits lower the opportunity cost of recuperating, delaying workers' return to work. Regardless of whether the delay in returning to work is medically valid, reported injury severity rises even if true severity falls. Worrall and Appel (1982) indirectly controlled for the true severity of injury by examining the impact of WC benefits on the ratio of the cost of indemnity claims to the cost of medical only claims. A rise in true severity should cause medical costs to expand at the same rate as indemnity payments, while a rise in reported severity should cause indemnity payments to expand faster than medical costs. Worrall and Appel's results demonstrate a significant reporting effect on injury severity.

Besides directly increasing benefits for income loss from injury, states can expand the generosity of their WC systems by reducing the waiting period to draw income benefits. To prevent workers from receiving wage loss compensation for relatively minor injuries most states require at least a three day absence from work, and many require at least a seven day absence, before drawing income benefits. Extending the length of the waiting period reduces the compensation for minor injuries, thereby encouraging workers to increase safety efforts. Consistent with economic incentives, Chelius (1982, 1983), Butler and Worrall (1983), and Krueger (1990a) found workplace injuries lower in states with relatively long waiting periods, all else equal. Based on Krueger's estimates expanding the length of the waiting period from three to seven days reduced the probability of receiving income benefits by 38.7 percent.

Experience Rating. From economic theory the more responsive WC premiums are to changing safety levels the more dramatically firms will expand their safety efforts as WC benefits rise. In practice, insurance companies combine injury statistics for all workers in an industrial class to determine the so-called manual rate for WC coverage. For small firms the product of each worker's wage rate and the manual rate for his or her industrial class summed over all workers determines the firm's total WC insurance premium. As firm size rises insurers pay less attention to the average experience of the industrial class (the manual rate) and more attention to the safety record of the individual firm. Experience rating refers to the process of adjusting insurance prices to reflect individual versus group experience. Currently, insurance providers completely experience rate WC premiums for only the largest establishments.

Because the degree of experience rating varies by firm size an increase in WC benefits should cause nonfatal injuries to rise less and fatal injuries to fall more in larger firms. Many of the studies summarized in Table 2-2 have attempted to control for firm size to determine the impact of experience rating on workplace safety. As expected, Ruser (1985, 1991) and Worrall and Butler (1988) found that when WC benefits increased large highly experience rated firms have had the smallest overall rise in workplace injuries. Likewise, Moore and Viscusi (1989, 1990) discovered higher benefits decreased workplace fatalities most in large highly experience rated firms. Only Chelius and Smith (1983) did not uncover a significant impact on safety from experience rating. Even here, though, the general direction of their results provide weak support for the belief that experience rating is safety enhancing.[10]

THE NEED FOR NUMERICAL SIMULATION RESEARCH

The conclusion emerging from the econometric studies of OSHA we have summarized is that workplace safety regulation, as currently operating in the United States, has not led to much improvement in workplace safety. The logical next step to researching OSHA's effects would be to consider the possibility of extreme changes in the OSHA's parameters, such as greatly increasing the rate of inspections or the level of fines for violating safety standards.

It is well known that the error of a forecast from a regression model grows with the distance of the exogenous variables from their mean values (Greene 1993, Chapter 6). The large variance that would be attached to an econometric prediction means that we are unwilling to trust regression equation estimates to forecast how extreme changes in OSHA may affect labor market outcomes. Moreover, U.S. data are insufficiently rich to allow inferring possibly subtle interactions between OSHA and states' workers' compensation insurance programs. The logical alternative research approach to econometric models of OSHA is to specify a complete model of the labor market heterogeneity of firms and workers based on our current empirical knowledge of structural parameters and then numerically simulate behavioral responses to possible policy reforms. We thus will use a numerical simulation model to identify whether extreme changes in OSHA, including huge fines and nearly perfect safety inspection, are reasonable policies to strengthen the workplace safety incentives OSHA's framers intended it to have.

The econometric evidence we have summarized is also cloudy concerning whether the income support WC provides injured workers leads on balance to more or less safe workplaces due to the conflicting incentives WC provides workers and firms. Again the data are insufficiently rich to separate workers' incentives to report a possibly false workplace injury from firms' incentives to provide safer workplaces via experience rated insurance premiums. We will therefore use a numerical simulation model to infer whether there can be further safety gains from a tighter linkage between a firm's injury history and its WC prices and to separate the true safety impact of WC from its incentives to report injuries. Our numerical simulations will help redirect future econometric research by locating which structural equations are the most important to estimate and identifying what are the most informative measures of WC's behavioral influences on workplace safety.

CONCLUSION

OSHA enforcement efforts directly and indirectly influence workplace safety. Firms directly respond to more frequent inspections and higher fines for noncompliance by expanding their safety efforts, thereby reducing accidents and diseases. Indirectly, the improvement in safety alters the relationship between wages and risk causing firms to reevaluate additional expenditures on safety. Although the indirect impact of OSHA mitigates to some extent its direct impact, only under extreme and fairly unrealistic situations will greater enforcement of workplace safety standards reduce workplace safety. Existing econometric evidence indicates that OSHA has had, at best, a modestly positive impact on workplace safety. In the next chapter we develop the numerical simulation model we use in Chapter 4 to estimate the safety effects of greatly increasing the level of OSHA inspections and fines and to discover any policy benefits from trying to coordinate the parameters of OSHA with the states' workers' compensation insurance programs.

WC insurance also produces direct and indirect impacts on safety but, unlike OSHA, WC directly influences both workers and firms. Higher WC benefits reduce workers' economic losses from injury, which lowers workers' concerns with safety. At the same time higher WC benefits also expand firms' interests in minimizing insurance expenses by reducing accidents and diseases. Changes in the relationship between wages and risk bring together the opposing interests of workers who desire less safety and firms who desire greater safety. The indirect impact of WC determines whether higher insurance benefits increase or decrease workplace injuries.

Econometric evidence indicates that pain and suffering are the largest cost to workers of an accident or a disease. By slightly reducing income loss, higher WC benefits should not greatly lower workers' desires to avoid unsafe or unhealthy work environments. On the other hand, the expanding costs of providing WC coverage as benefits rise should cause firms to find means to improve safety. In equilibrium, firms' incentives should dominate workers' incentives, and safety should improve. When researchers have examined injury rate measures relatively free of reporting effects, they discovered that WC dramatically improves workplace safety. Many studies failed to discover the resulting improvement in safety, however, because higher WC benefits encourage some workers to file claims for injuries they would previously have not reported because it was not worth the effort and encourages some other workers to file WC claims for specious or minor accidents and diseases. The numerical simulation model we develop in the next chapter is also the research tool we use in Chapter 5 to identify quantitatively how the link between the incentives to file for WC benefits causes reported injuries to vary substantially from real injuries.

NOTES

[1]Injury rates in construction were 13.1 per 100 workers versus 2.9 per 100 workers in finance, insurance, and real estate; fatality rates in construction were 14 per 100,000 workers versus 2 per 100,000 workers in finance, insurance, and real estate (BLS 1993 and Tascano and Windau 1993).

[2]It is unclear whether safety programs increase or decrease production. Viscusi (1979) argues safety equipment increases output by diminishing the disruptive effects of injuries and by increasing the stability of the workforce. However, programs such as slowing the pace of the assembly line or installing cumbersome machine guards can interfere with the work process and decrease output. On net, which effect of safety equipment dominates is currently an unresolved empirical question. We assume safety equipment is a productive factor when developing our numerical hedonic equilibrium model, which gives OSHA the benefit of the doubt when simulating the effects of safety regulations.

[3]Although wages fall for all workers, a reduction in the number of deaths means a greater proportion of workers are paid for the full period. The first effect reduces and the second effect increases the firm's wage bill. We assume for the purpose of discussion that the wage reduction effect swamps the wage bill increase from fewer accidents.

[4]Representative William Steiger, a principal author of the Occupational Safety and Health Act, claimed injuries would fall by, "50 percent or something like that" (Viscusi 1992, p. 290).

[5]Worker welfare rises if the new wage function intersects the original equilibrium acceptance wage (Θ_1 in Figure 2.4).

[6]The impact might be considerably smaller considering that many studies found neither abatement nor deterrence effects from OSHA inspections.

[7]Safety may also improve for three other reasons. First, WC increases the cost of labor relative to capital, thereby reducing employment and workplace injuries. Second, if WC benefits are related to pre-injury wages then firms improve safety to minimize the cost of purchasing the insurance necessary to cover the wage compensation for risk. Specifically, if $b=iw$ where i is a constant exceeding zero, equation (2.6) (the condition for the firm's optimal level of safety) becomes

$$\frac{\partial R}{\partial S}\frac{\partial S}{\partial \pi} - \left(\frac{\partial W}{\partial \pi}(1-\pi) - w\right)n - \frac{\partial P_b}{\partial \pi} iwn - P_b\frac{\partial W}{\partial \pi} in - \frac{\partial V}{\partial \pi} = p_s\frac{\partial S}{\partial \pi}.$$

As the level of benefits, i, expand firms reduce workplace hazards to lower the per-unit price of insurance $\left(\frac{\partial P_b}{\partial \pi} iwn\right)$ and the required amount of insurance coverage $\left(P_b \frac{\partial W}{\partial \pi} in\right)$. Finally, WC may increase safety because insurance prices vary across industries. Even if the cost of WC insurance is the same for all firms within an industry, firms in high risk industries pay more per-unit of insurance coverage than firms in low risk industries; therefore, an increase in WC benefits raises employment costs more in high risk industries than in low risk industries, all else equal. The number of workplace injuries falls as workers move from hazardous to safe industries (Butler 1994).

[8]With workplace fatalities relatively rare, any measure of total injuries is primarily a measure of nonfatal injuries.

[9]The degree of mismeasurement should also differ by the type of nonfatal injury. Reporting effects should be greatest for injuries where the existence or the extent of impairment is difficult to verify medically, such as sprains and strains, and least for easily diagnosed injuries, such as lacerations and fractures. Consistent with a variable reporting effect, Robertson and Keeve (1983) found higher WC benefits increased the number sprains and strains more dramatically than the number of lacerations and fractures.

[10]The results in Ruser (1993) paint a mixed picture of the impact of experience rating. The fatal injury rate/benefit elasticities do not monotonically rise (in absolute value) as establishment size grows, a result not supportive of an experience rating effect. Nevertheless, consistent with large firms having greater incentives than small firms to reduce injuries as benefits rise the fatal injury rate/benefit elasticity for the top employment size group is always strongly significant across all empirical specifications whereas the elasticities for two of the three smaller size groups are always insignificant.

3 THE SIMULATION MODEL

Economists since Adam Smith have been interested in hedonic equilibrium, or how the characteristics of a good or service are related to its price. The interest encompasses both product markets, as typified by the automobile industry study of Griliches (1971), and labor markets, as in the study of worker safety and compensating wage differentials by Thaler and Rosen (1975). Econometric research summarized in Chapter 2 reveals the difficulty of estimating the hedonic equilibrium price function (Brown and Rosen 1982, Brown 1983, Epple 1987, and Kahn and Lang 1988). Theoretical research also demonstrates the impossibility of deriving an analytical solution for the hedonic equilibrium locus in the presence of uncertainty caused, for instance, by the possibility of a work-related accident or disease (Epple, p. 63). To overcome the limitations of existing econometric and theoretical research we developed a numerical simulation model capable of examining both the direct and the indirect effects of public policies geared toward improving workplace safety.

Numerical simulation is a valuable research tool for several reasons. First, in considering the complete set of interactions among workers, firms, and the labor market, most comparative static results concerning safety effects are ambiguous in sign, a priori. More importantly, existing data are insufficient to examine completely new programs of interest such as imposing an injury tax on employers, or to examine extreme changes in existing programs such as massive increases in OSHA enforcement efforts or substantial rises in WC benefits. We now provide a blueprint for researchers interested in numerically simulating how the labor market as tempered by government policy determines workplace safety.

The simulation model we now develop is the organizing framework for the remainder of the book. We begin by discussing two broad approaches for computing hedonic equilibria. We then detail the assumptions necessary to specify the parameters of the numerical model. Finally, we examine the sensitivity of the calculated equilibrium to changes in the model's background assumptions. The two convenient numerical simulation approaches we examined produced similar results. Likewise, moderate changes in starting parameter values produced only minor changes in the calculated equilibrium.

A MATHEMATICAL DESCRIPTION OF HEDONIC LABOR MARKET EQUILIBRIUM

The extremely high costs to remove or reduce many work-related health hazards prevents most firms from providing completely safe work environments. As shown in Chapter 2, profit maximizing firms determine their expenditures on safety enhancing measures by comparing the costs of additional safety to the benefits from reduced injury rates. Although only the direct costs may be readily apparent the benefits of safety programs may be substantial. For many firms reducing workplace hazards can raise profits by diminishing the disruptive effects of injuries, lessening labor turnover, reducing wage premiums, and decreasing the costs of WC insurance coverage. Similar to firms, workers must decide whether the additional wages received from relatively high-risk jobs compensate for the greater chance of injury. Because employers differ in their abilities to reduce injury risk and workers differ in their attitudes regarding accepting risk, no single probability of a work-related injury is optimal for all parties. Instead, workers and firms sort among various levels of injury risk (safety). In long-run equilibrium the relationship between wages and the probability of injury adjusts to equate the quantity of labor supplied to the quantity demanded at or near a given level of injury risk (Rosen 1974, 1986).

The hedonic model of labor market equilibrium transforms exogenous distributions of worker and firm characteristics into a common endogenous set of injury probabilities. The next two sections describe the transformation and resulting hedonic wage function. First we consider the situation where workers and firms can choose among many workplaces that differ by the risk of work-related death or total disability. Second we consider the possibility that safety enhancing investments are lumpy so that only a limited number of workplace types exist. Although the two situations are identical in the limit the mathematical derivations of the hedonic relationship are distinct and some notable differences emerged when workers can choose among many versus few employment opportunities.

Continuous Variation of Workplace Safety Levels

If safety levels are continuous then workers sort into a given job type (injury risk) based on the market (hedonic) wage function and their aversion to risk. Firms supply a given type of workplace based on the market wage function and their ability to produce safety. A shortage of workers in high risk establishments, for instance, will drive up wages, thereby enticing some workers away from relatively safe employment. At the same time the wage hike will encourage some firms to expand their expenditures on safety to reduce labor costs. With workers moving toward greater risk and firms moving toward less risk wages must rise in relatively safe establishments. An excess demand for labor at any point along the injury risk spectrum alters the delicate balancing of labor supply and demand everywhere. Wages adjust until the supply of labor equals the demand for labor along the entire injury risk spectrum.

Workers. Let α represent a worker's degree of risk aversion. All else equal the larger the value of α the more steeply sloped a worker's acceptance wage function (indifference curve). As we described in Chapter 2 workers most averse to risk sort toward relatively safe employment while workers less averse to risk sort toward relatively dangerous employment. Figure 3.1 illustrates a prototypical hedonic wage function. Andrew, who has an extremely high desire to avoid job hazards (as shown by Θ^A), maximizes expected utility by choosing risk level π^A and Emily, who has a lower need for safety (as shown by Θ^E), chooses risk level π^E.

Firms. Likewise, let μ represent a firm's ability to use safety equipment in the production of output. To maintain a constant level of profits firms with highly productive safety equipment require a lower wage sacrifice from their workers for a given improvement in safety than firms with less productive safety equipment. The larger the value of μ the flatter a firm's offer wage function (isoprofit curve). Because of differences in safety productivity some firms provide relatively risk free work environments while others expose their workers to extreme hazards. In Figure 3.1 Acme Electronics (Φ^A) maximizes profits by eliminating almost all workplace hazards while Edwardsville Chemical

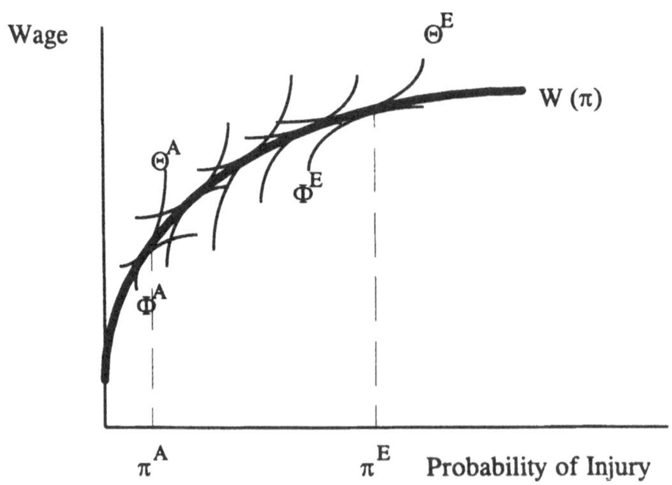

Figure 3.1 Labor Market Equilibrium

Manufacturers (Φ^E) maximizes profits by eliminating few hazards (so that $\pi^E > \pi^A$) but paying much higher wages.

Equilibrium. The hedonic wage function balances the supply and demand for labor along the entire risk spectrum. With our assumptions the equilibrium wage function must satisfy the following condition (Rosen 1974),

$$L(\alpha)\left|\frac{d\alpha}{d\pi}\right|d\pi = N(\mu)F(\mu)\left|\frac{d\mu}{d\pi}\right|d\pi, \tag{3.1}$$

where $L(\cdot)$ ≡ the density function of workers with respect to α, with $\alpha_{min} \leq \alpha \leq \alpha_{max}$,
$N(\cdot)$ ≡ the demand for labor by an individual firm, and
$F(\cdot)$ ≡ the density function of firms with respect to μ, with $\mu_{min} \leq \mu \leq \mu_{max}$.

The differentials $d\alpha/d\pi$ and $d\mu/d\pi$ depict the sorting of workers and firms into levels of risk — the mapping of acceptance and offer wage functions along the market wage function in Figure 3.1. The left-hand side of equation (3.1) represents the number of workers supplying labor to firms with a given level of safety while the right side represents the demand for workers. Because labor supplied and demanded must be identical in equilibrium we can use equation (3.1) to describe the equilibrium sorting of workers by the risk of a work-related injury. Further, by totally differentiating the first-order conditions for the maximization of expected profit with respect to μ (treating μ as an endogenous variable) we can develop an expression for the equilibrium sorting of firm characteristics and injury risk (see equations 2.4–2.6). Finally, we can determine the increase in wages necessary for workers to accept a given probability of injury ($dw/d\pi$) using the first-order conditions for expected utility maximization (see equation 2.2). A system of three simultaneous first-order differential equations completely describing hedonic equilibrium in the labor market is

$$\frac{d\alpha}{d\pi} = A'(\pi, \alpha, \mu, w, M'), \tag{3.2}$$

$$\frac{d\mu}{d\pi} = M'(\pi, \alpha, \mu, w, A', W'), \text{ and} \tag{3.3}$$

$$\frac{dw}{d\pi} = W'(\pi, \alpha, w). \tag{3.4}$$

The details of the hedonic equilibrium system, including the relevant boundary conditions, are described in Appendix 3A. In hedonic equilibrium the first-order conditions for utility and profit maximization are satisfied and the labor market is in equilibrium at all levels of workplace safety. The three differential equations (3.2), (3.3), and (3.4) determine the relationship between worker characteristics and injury risk, firm characteristics and risk, and the hedonic wage function.

Discrete Variation of Workplace Safety Levels

In an alternative form of the hedonic labor market equilibrium model we consider the situation where purchases of safety equipment are necessarily lumpy, perhaps caused by a limit on the types or quantities of safety equipment available, which restricts the diversity of workplace safety. If only a few safety levels exist, each corresponding to one of the possible discrete magnitudes of safety equipment, then we can no longer use marginality conditions to study the behavior of workers or firms. Instead, we must view market participants as maximizing welfare by ranking each safety submarket according to the level of expected utility or profit and then choosing the submarket with the highest ranking. We retain all other assumptions from the previous section. In the discrete version of the hedonic equilibrium model wages adjust to equilibrate the supply and demand for labor within each of the limited number of different injury rate submarkets.

Workers. On the supply side of the labor market a person seeks the level of safety maximizing his or her expected utility. The selection process sorts people into the various submarkets in light of their heterogeneity with respect to safety preferences. Let $\overline{U}(\cdot)$ represent a worker's indirect expected utility function. A worker chooses the submarket (safety level) associated with i and rejects employment in submarket $i+1$ if

$$\overline{U}(w_i,\pi_i;y,\alpha) - \overline{U}(w_{i+1},\pi_{i+1};y,\alpha) > 0 \tag{3.5}$$

where submarkets are ordered from 1 to j, with submarket 1 characterized by the greatest level of workplace safety and submarket j by the lowest level of safety.

Now define α_i as the value of α where equation (3.5) holds with strict equality. Assuming α, the individual's risk aversion parameter, is continuously distributed workers with risk aversion greater than α_i choose employment with levels of safety greater than or equal to that in submarket i, which is $(1-\pi_i)$, while workers with lower risk aversion accept greater health hazards.[1] Looking across all submarkets, we can identify j-1 values of α that uniquely sort workers into the j submarkets.

Given the rational sorting of workers just described, where the α values depend on all submarket wage rates and injury rates the supply of labor to each submarket is

$$\begin{aligned} n_1 &= \int_{\alpha_1}^{\alpha_{max}} L(\alpha)\,d\alpha, \\ n_2 &= \int_{\alpha_2}^{\alpha_1} L(\alpha)\,d\alpha, \\ &\vdots \\ n_j &= \int_{\alpha_{min}}^{\alpha_{j-1}} L(\alpha)\,d\alpha. \end{aligned} \tag{3.6}$$

Firms. Where the demand for labor is concerned firms make joint decisions on the use of labor, capital, and safety equipment. As when discussing workers' decisions on work-related injury risk when only a limited number of safety levels are possible we cannot use marginality conditions to determine firms' optimal purchases of safety equipment.

Instead, firms first compare expected profit at each possible level of safety, optimizing their use of labor and capital, then choose the most profitable levels of inputs.

Let $G(\cdot)$ be the firm's indirect profit function. A firm chooses to provide the safety level implied by π_i as opposed to π_{i+1} if

$$G(w_i, \pi_i; \mu) - G(w_{i+1}, \pi_{i+1}; \mu) > 0. \tag{3.7}$$

Now define μ_i as the value of μ making (3.7) hold as an exact equality. Assuming that μ, the firm's safety equipment efficiency parameter, is continuously distributed firms with μ greater than μ_i provide levels of workplace safety greater than or equal to $1\text{-}\pi_i$ while firms with μ lower than μ_i provide less safety.[2] Looking across all levels of safety we can identify j-1 values of μ that sort firms into the j possible submarkets with each value depending on all wage rates, injury rates, and parameters of the expected profit function.

By using the endogenously determined distribution of firms described in the previous paragraph to aggregate individual buyers' demands for labor we can determine the total demand for labor within each safety submarket. Expressed algebraically the total demand for labor at each level of safety is

$$\begin{aligned} n_1^d &= \int_{\mu_1}^{\mu_{\max}} N_1(w_1; \mu) F(\mu) d\mu, \\ n_2^d &= \int_{\mu_2}^{\mu_1} N_2(w_2; \mu) F(\mu) d\mu, \\ &\vdots \\ n_j^d &= \int_{\mu_{\min}}^{\mu_{j-1}} N_j(w_j; \mu) F(\mu) d\mu, \end{aligned} \tag{3.8}$$

where $N_i(w_i; \mu) \equiv$ the optimal use of labor by an individual firm providing the ith level of safety $(1\text{-}\pi_i)$; $N(\cdot)$ is determined by the simultaneous solution of the first-order conditions with respect to labor and capital.[3]

Equilibrium. General equilibrium in the labor market occurs when a set of j wage rates equalizes the quantity of labor supplied with the quantity demanded at each of the j possible levels of safety. We can determine wages by equating each equation in (3.6) to the corresponding equation in (3.8) and simultaneously solving for the j endogenous wages.

Figure 3.2 shows the equilibrium situation with three safety levels: low, moderate, and high ($\pi_3 > \pi_2 > \pi_1$). The key to understanding the conditions for equilibrium in the discrete version of the hedonic equilibrium model is realizing the three submarkets are interconnected. An excess demand for labor in the low safety submarket, for instance, causes some high-risk firms to raise wages to attract workers away from firms offering greater safety. At the same time, other high-risk firms will increase their safety investments to reduce workplace hazards to capture the now relatively greater wage reduction from improved safety. The movement of workers toward less safety and firms toward greater safety, in turn, produces an excess demand for labor in the other two submarkets. In short, any wage change in one safety submarket soon causes wage changes in every other safety submarket. Complete equilibrium requires a balancing of supply and demand throughout the injury risk spectrum.

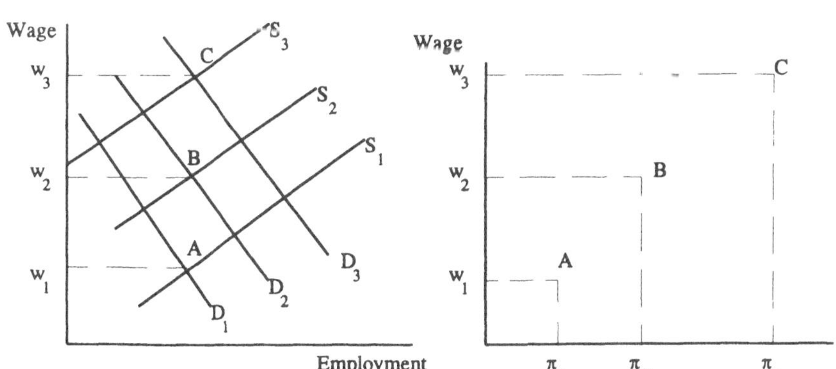

Figure 3.2 Equilibrium with Limited Safety Levels

NUMERICAL SIMULATION TECHNIQUES

We will demonstrate two methods to generate the equilibrium wage function numerically. The first employs a Taylor series to approximate the system of differential equations representing equilibrium in the continuous version of the hedonic model. The second solves the set of simultaneous equations comprising the discrete version of the model. In the limit the two methods are identical. By decreasing the interval between safety levels, or by increasing the number of submarkets, the solutions should theoretically approach a common value. Nevertheless, the economic interpretations of the two numerical techniques differ. One model mathematically approximates the true continuous economic system generating the hedonic wage function, while the other model approximates the true continuous system with the economist's favorite tool — supply and demand. The continuous outcomes model assumes workers and firms mass only at the probability of injury representing complete workplace safety ($\pi = 0$), while the discrete outcomes model assumes workers and firms mass at every possible level of safety.

A Taylor-Series Approximation

The continuous hedonic equilibrium model is fully described by three simultaneous first-order differential equations and associated boundary conditions. The boundary conditions establish equilibrium at the zero and maximum risk levels by adjusting the value of wages and matching the least risk-averse workers to firms who are the least efficient producers of workplace safety. The zero profit condition determines the maximum level of injury risk.

Initial conditions defined at two points along the risk spectrum create so-called boundary value problems in that the solutions to the differential equations depend on the initial conditions but the initial conditions depend on the solutions to the differential equations. The least complicated method of solving boundary value problems is to approximate the functions $A(\pi)$, $M(\pi)$, and $W(\pi)$ in equations (3.2), (3.3), and (3.4) with first-order Taylor series. Specifically, the functions for differences in risk preferences, safety efficiency, and wages across workers and firms can be represented as the system of 3(*j*-1) simultaneous nonlinear equations

$$\alpha_{i+1} = \alpha_i + hA'(\pi_i, \alpha_i, \mu_i, w_i, M_i'), \qquad \textbf{(3.9)}$$

$$\mu_{i+1} = \mu_i + hM'(\pi_i, \alpha_i, \mu_i, w_i, A_i', W_i'), \text{ and} \qquad \textbf{(3.10)}$$

$$w_{i+1} = w_i + hW'(\pi_i, \alpha_i, w_i), \qquad \textbf{(3.11)}$$

where $h \equiv \pi_{i+1} - \pi_i$,
$i \equiv 1, \cdots, j-1$, $\pi_1 = 0$ and $\pi_j = \pi^m$, and
A', M', and W' represent the differential equations defined previously. (Subscripts indicate functions evaluated at $\pi = \pi_k$, $k = 1, \cdots j$; and π^m is the maximum probability of injury.)

Combining the equations in (3.9)–(3.11) with the boundary conditions establishes a system that simultaneously solves for the equilibrium values of the endogenous variables at various probabilities of injury, including zero and the maximum injury probability. Numerical accuracy increases as the interval of differencing decreases (as $h \to 0$), assuming round-off error is small. In the reported simulations we decreased the interval of differencing until all boundary solutions changed by less than 0.1 percent.[4]

Using Supply and Demand Relationships

The discrete version of the hedonic equilibrium model is the basis of the second numerical technique we used to generate the relationship between wages and injury risk. Because the system of nonlinear equations in (3.6) and (3.8) theoretically represents equilibrium completely the technique does not encounter boundary value problems. In addition, as the number of submarkets increases the model approaches the continuous case so numerical differences between the two approaches to numerical simulation should be minor. Integrating labor demand and supply within each submarket creates the greatest practical problem in implementing the discrete approach. In general, functions with widely varying solution values over the range of integration require sophisticated numerical techniques to reduce approximation error. In our case the assumptions we make concerning the initial parameter values for worker and firm heterogeneity, α and μ, and the number of submarkets we use allow us to use a relatively simple numerical integration method, Simpson's rule, without sacrificing accuracy.

PARAMETERIZING THE SIMULATION MODEL

Any method used to solve for the hedonic wage function requires specific functions to represent (i) workers' preferences, (ii) firms' technology, and (iii) interperson and interfirm differences in α and μ (risk aversion and efficiency of safety measures). In choosing parameters we attempted to reproduce the pre-OSHA labor market outcomes studied by Thaler and Rosen (1975). From the base case we then investigated how changes in the underlying parameters influence the hedonic equilibrium wage function and in subsequent chapters identify some of the economic implications of public policies geared to increasing workplace safety. Our choice of structural equations and initial parameter values are summarized in Table 3-1 and justified more fully in Appendix 3B.

Figure 3.3 outlines the general approach used to formalize the mathematical model.[5] As with studies relying on computable general equilibrium we began by specifying a core set of endogenous values the model should reproduce to represent accurately the economy during some base year. In our situation we wish to reproduce the labor market at the end of the 1960s, a period before the major policy initiatives aimed at improving industrial safety such as the Occupational Safety and Health Act and the National Commission on State Workmen's Compensation Laws.

The initial conditions represent the core set of values from which policy changes are evaluated. After specifying the initial situation we

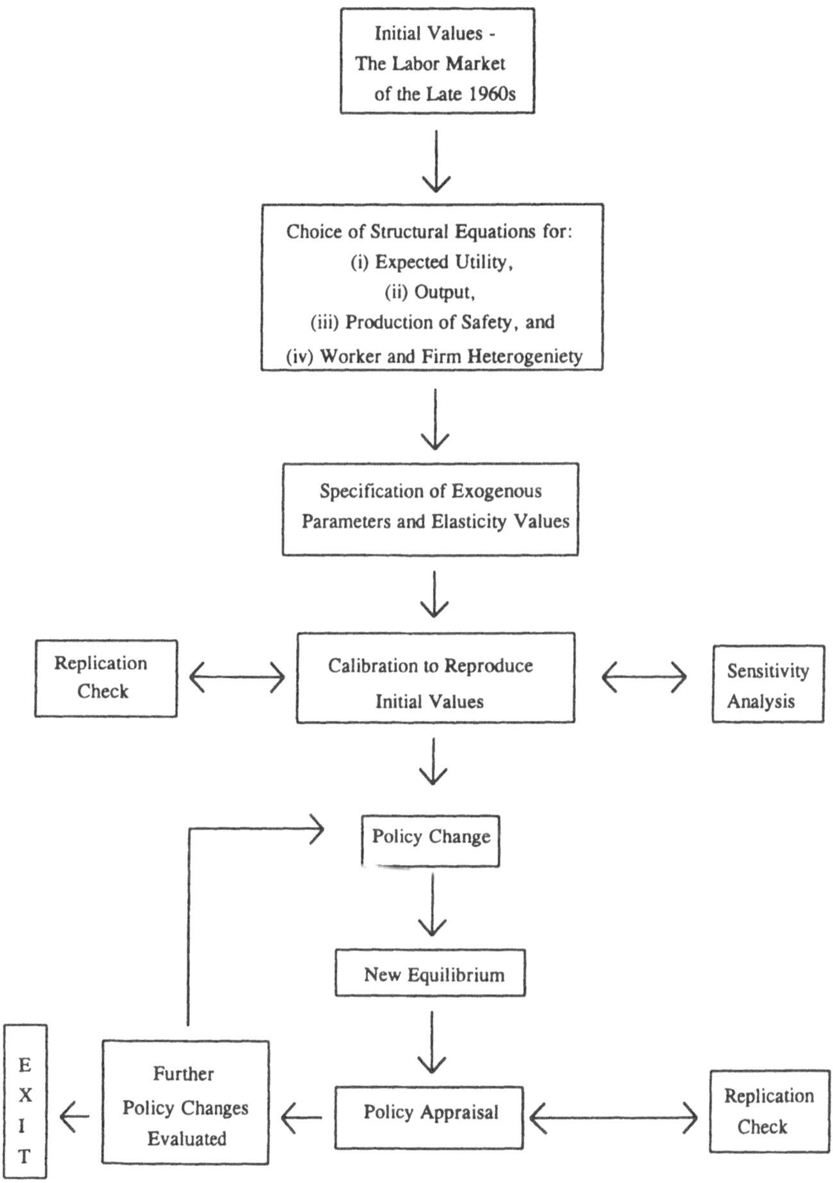

Figure 3.3 Calibrating the Numerical Model

Table 3-1. Base Case Functions and Starting Parameter Values

I. FUNCTIONS

Expected utility:

$$\bar{u} = (1-\pi)(w+y)^{1-\alpha} + \pi\beta(b+y)^{1-\alpha}$$

Production of output:

$$q = a\left[\delta(\mu s)^{-\rho} + \gamma k^{-\rho} + (1-\delta-\gamma)n^{-\rho}\right]^{-\upsilon/\rho}$$

Production of safety:

$$s = s_0 + s_1\pi + s_2\pi^2$$

II. WORKERS

Variable:	α	β	Efficiency units of labor per worker	y	b
Value:	[0.96, 0.98]	0.1	10,000	0	0

III. FIRMS

Variable:	μ	a	δ	γ	υ	ρ	ν	c
Value:	[0, 1]	Varies	0.10	0.20	0.1	0.01	0	Varies

DEFINITIONS

a	≡	the efficiency parameter
b	≡	insurance benefits if injured
c	≡	fixed costs
k	≡	capital
$M(0)$	≡	the value of μ for the last firm locating at $\pi = 0$
$M(\pi^m)$	≡	the value of μ for the firm locating at the maximum level of injury risk
n	≡	labor
p_k	≡	the price per unit of capital
p_o	≡	the price of output
p_s	≡	the price per unit of safety equipment
q	≡	output
s	≡	safety equipment
$\bar{u}$	≡	expected utility
ν	≡	the expected fine for safety and health violations
w	≡	the wage rate
$W(0)$	≡	the wage rate in perfectly safe establishments

IV. MARKETS

Variable:	p_k	p_s	η	Number of workers	Number of establishments
Value:	\$1	\$1	-1	1.52427×10^7	311,140

V. DISTRIBUTIONS

Workers are distributed uniformly with respect to α using the density function:

$$L(\alpha) = 7.621 \times 10^8 \mid 0.96 \leq \alpha \leq 0.98,$$
$$L(\alpha) = 0 \text{ elsewhere.}$$

Firms are distributed uniformly with respect to μ using the density function:

$$F(\mu) = 3.1114 \times 10^5 \mid 0 \leq \mu \leq 1,$$
$$F(\mu) = 0 \text{ elsewhere.}$$

VI. INITIAL CONDITIONS

$W(0)$ = \$10,000 (\$37,242, 1993 dollars)	p_0 = \$1
Average injury rate = 0.001	$\pi^m = 0.003$
$M(0) = 0.64$	$M(\pi^m) = 0.1$

y	$\equiv$	nonlabor income
α	$\equiv$	the degree of risk aversion parameter
β	$\equiv$	the bequest motive parameter
δ, γ	$\equiv$	distribution parameters
η	$\equiv$	the product price elasticity
μ	$\equiv$	the efficiency of safety equipment in the production of output
υ	$\equiv$	the returns to scale parameter
π	$\equiv$	the probability of a work-related injury resulting in total disability or death
π^m	$\equiv$	the maximum level of injury risk
ρ	$\equiv$	the elasticity of substitution parameter, $\sigma = 1/(1+\rho)$

then chose specific structural equations to represent expected utility, production of output, production of safety, and distributions of workers and firms with respect to risk aversion and safety equipment productivity. We used the simplest functional forms consistent with the theory developed earlier in the chapter. Based on the structural equations we then specified as many parameter values as possible based on existing econometric research.

Calibration

As normally happens in computable general equilibrium models one cannot determine every parameter value by relying solely on existing econometric estimates. In particular, little information is currently available concerning the relationship between safety programs and workplace safety, which is the safety production function embedded in the expected profit function (see equation 2.3). As a result we calibrated the model using the safety production function. Calibration means adjusting a parameter or a set of parameters so the numerical model reproduces a core set of values (Mansur and Whalley 1984). In choosing initial values we reproduced the situation covered by the 1967 Occupational Study of the Society of Actuaries. The Occupational Study provides an excellent source of data on the incremental probability of death from both industrial accidents and diseases (Thaler and Rosen 1975) and does not suffer from the under reporting that prevails in establishment data (Burroughs and Lubove, 1986). Still, the Society of Actuaries study is not without faults; it may overstate the true frequency of injury because the data cover only hazardous occupations and because people who choose high-risk occupations may also accept greater risk off the job. Although arguments can be made in favor of other sources of injury rate data we chose the Occupational Study to incorporate the risk of industrial disease resulting in death — a fatality that is largely ignored in other injury rate statistics. In the sample studied by the Society of Actuaries upon which our simulations are based the mean incremental probability of a work-related death is 1 in 1,000.

We assumed a quadratic safety production function, $s = s_0 + s_1\pi + s_2\pi^2$, meaning we had to determine the constant, s_0, and the coefficients on risk and its square, s_1 and s_2. Hence, to calibrate the model completely we needed two additional conditions. Based on data reported in Oi (1973, p. 72) we adopted 0.4 as the proportion of firms without incremental workplace hazards which implies in our model a value of μ for the last firm located at $\pi = 0$, denoted $M(0)$. We also assumed 0.003 is the highest probability of injury observed in the labor market (π^m), a value roughly corresponding to the greatest occupational risk category reported in Thaler and Rosen. Finally, as discussed more fully in Appendix 3B the other three initial conditions — the price of output equaling \$1, the zero risk wage equaling \$10,000, and the value of μ for the most hazardous firm equaling 0.1 — allowed us to determine the technical efficiency parameter (a), the level of fixed costs (c), and the coefficient for the constant elasticity product demand function.[6]

Validity Checks

Once the model was fully calibrated, we determined whether the initial equilibrium replicates the starting values used in the calibration process, a step that is a method of checking for errors in the computer program. Because our selection of parameter values and initial equilibrium conditions may crucially affect the calculated hedonic equilibrium, we also did extensive sensitivity checks before using the numerical model to study changes in public policy. Our public policy evaluation follows the standard lines used in computable general equilibrium. We first incorporated a proposed policy change into the mathematical model and recomputed equilibrium. We then compared the initial equilibrium to the new equilibrium to determine net changes in safety or other measures of well-being.

In some situations econometric research already details the impact of existing programs geared toward improving workplace safety. Specifically, we know that OSHA regulations have generally not improved workplace safety and that increased WC insurance benefits have raised nonfatal injuries but lowered workplace fatalities (see Chapter 2). As a check of the validity of our numerical simulation model we compared our results to existing econometric estimates to determine how closely the model replicates what we already know about the basic properties of OSHA and WC. The ability to replicate what we already know econometrically about job safety policy improves our confidence in the model's ability to uncover how some basic changes in public policy might affect workplace safety.

INITIAL NUMERICAL SIMULATION RESULTS

Table 3–2 presents the values of the initial equilibrium wage function associated with both the continuous and the discrete versions of the numerical model. As explained previously the model has been calibrated to produce a $10,000 rate of pay in perfectly safe establishments and a 1 in 1,000 average frequency of a work-related injury resulting in death or total disability, values roughly reflecting labor market conditions in the United States in the late 1960s and early 1970s. To account for the impact of inflation to 1993 all monetary figures have been adjusted using the consumer price index. In terms of constant 1993 dollars the maximum absolute difference between the continuous and discrete formulations is only $34, or 0.08 percent. The most striking feature of Table 3–2 is the similarity of the calculated wages across simulation techniques.

Acceptance and Offer Wages

Also included in Table 3–2 are the acceptance wage function for the most risk-averse (α = 0.98) and the least risk-averse (α = 0.96) workers and the offer wage function for firms with the most productive (μ =1) and least productive (μ = 0.1) safety equipment. Acceptance and offer wage functions represent the monetized value of the indifference and isoprofit curves in wage/risk space. We derived asking and offer wages using the level of utility or profit for a completely safe

Table 3-2. The Initial Computed Hedonic Wage Function (1993 dollars)

Technique/Concept	Wage $\pi =$ 0	Wage $\pi =$ 0.00025	Wage $\pi =$ 0.0005	Wage $\pi =$ 0.00075	Wage $\pi =$ 0.001	Wage $\pi =$ 0.00125	Wage $\pi =$ 0.0015	Wage $\pi =$ 0.00175	Wage $\pi =$ 0.002	Wage $\pi =$ 0.00225	Wage $\pi =$ 0.0025	Wage $\pi =$ 0.00275	Wage $\pi =$ 0.003
DIF	$37,242	$37,570	$37,894	$38,214	$38,531	$38,840	$39,142	$39,440	$39,734	$40,017	$40,296	$40,568	$40,829
S&D	37,242	37,574	37,398	38,214	38,527	38,836	39,134	39,432	39,715	39,998	40,270	40,534	40,795
Acceptance Wage $\alpha = 0.98$	37,242	37,712	38,184	38,665	39,153	39,644	40,143	40,650	41,164	41,682	42,207	42,739	43,279
Acceptance Wage $\alpha = 0.96$	37,242	37,477	37,708	37,946	38,184	38,427	38,665	38,911	39,153	39,399	39,644	39,898	40,147
Offer Wage $\mu = 1$	37,242	37,570	37,894	38,211	38,523	38,829	39,130	39,421	39,708	39,987	40,255	40,516	40,769
Offer Wage $\mu = 0.1$	37,242	37,581	37,913	38,240	38,564	38,881	39,190	39,492	39,786	40,073	40,352	40,624	40,885
Comp	1,315	1,296	1,277	1,445	1,236	1,214	1,192	1,166	1,140	1,114	1,084	1,054	

Note: DIF ≡ the simulation technique directly solving the three differential equations comprising the continuous model.

S&D ≡ the simulation technique based on the submarket supply and demand for labor.

Acceptance Wage ≡ the shadow value of the iso-utility function for workers with risk aversion given by a particular value of α and $\pi = 0$.

Offer Wage ≡ the iso-profit function value for firms with safety productivity given by a particular value of μ and $\pi = 0$.

Comp ≡ the slope of the wage function evaluated for a 1/1000 increase in the probability of a work-related injury.

workplace. To produce the offer wage function we used the safety production function generated by calibrating the continuous version of the model. By comparing the acceptance and offer wage functions with the hedonic equilibrium wage function we can determine how closely labor market equilibrium mirrors either the tastes and preferences of workers or the technology of firms.

Considering the narrowness of risk preferences ($0.96 \leq \alpha \leq 0.98$) one might anticipate the hedonic equilibrium wage function would closely match workers' indifference curves. At the maximum level of risk the hedonic equilibrium wage is \$2,450 to \$2,484 (approximately 6 percent) lower than the acceptance wage for the most risk-averse worker and \$648 to \$682 (approximately 2 percent) greater than the acceptance wage for the least risk-averse worker. Despite the seemingly small amount of worker heterogeneity there are still meaningful differences among the extreme values for the acceptance wage functions and either of the calculated hedonic equilibrium wage relationships.

Although we defined the efficiency parameter of safety equipment over a much larger range of values ($0 \leq \mu \leq 1$) than the risk-aversion parameter the hedonic equilibrium relationship more closely represents the underlying technology of firms than the preferences of workers. The equilibrium wage at $\pi = 0.003$, the maximum risk level, is only \$26 to \$60 (approximately 0.1 percent) greater than the offer wage for the firm with the most productive safety equipment, and \$56 to \$90 (approximately 0.2 percent) lower than the firm with the least productive safety equipment.

We also used the acceptance and offer wage functions to check whether the calculated hedonic wage function in Table 3–2 is an equilibrium wage function. In equilibrium, relatively risk-averse workers and relatively efficient producers of safe workplaces should sort into the lower levels of risk, and workers who care less about risk and firms who are less efficient at creating a safe workplace should sort into the highest levels of risk. The hedonic equilibrium sorting implies the acceptance wage for $\alpha = 0.98$ and the offer wage for $\mu = 0.1$ should both lie above the hedonic equilibrium function, and the acceptance wage for $\alpha = 0.96$ and the offer wage for $\mu = 1$ should both lie below the hedonic equilibrium function. If this were not so then the most risk-averse workers and the most productive producers of a safe workplace could increase economic well-being by accepting greater risk. As an additional check we examined the second-order conditions. In every instance the calculated hedonic wage function satisfies the sufficient conditions for utility and profit maximization.

Risk Compensation

Table 3–2 also presents calculations for risk compensation using the continuous version of the model. Results for the discrete version are comparable. Risk compensation is the slope of the wage function evaluated for a 1 in 1,000 increase in the probability of a work-related injury. Risk compensation ranges from $1,054 to $1,315 or from 3 to 4 percent of wages in Table 3–2. Thaler and Rosen (1975) estimated that approximately a 3 percent increase in wages is needed to get a worker to accept a 0.001 incremental probability of death. In his review article Smith (1979) reported percentage differentials in risk compensation across studies ranging from 1 to 20 percent, with the majority in the 2 to 4 percent range. Thus, our numerical simulations produce a compensation for the risk of a work-related injury that corresponds well with estimates in the empirical literature.

How risk compensation varies with risk levels, or the shape of the hedonic equilibrium locus is open to some debate. Smith (1979, p. 340) argued it must be concave from below because of increasing marginal costs of supplying safety across firms. Viscusi (1978, p. 383–374) argued similarly, but said workers' risk preferences determine the shape of the function. Because workers least averse to risk will accept a lower compensation per unit of risk they sort into high-risk jobs with lower wage premiums per unit of risk. Thaler and Rosen (1975) made no claim concerning the second derivative of the hedonic wage function but graphically depict wages increasing at an increasing rate. Our simulations in Table 3–2 also help us understand the shape of the hedonic wage locus.

Theoretically, the concavity of the wage function jointly depends on the distributions of safety technologies and risk preferences across firms and workers. If providing safety is extremely costly for some firms and few relatively risk-preferring workers exist, then the equilibrium hedonic wage function will be convex from below. Otherwise, the wage function may be either concave or linear. As can be seen in Figure 3.4, with the parameter values and distributions chosen we calculated a concave hedonic equilibrium wage function but the curvature is not dramatic. The compensation for risk declines over the risk spectrum by $261, or 20 percent. Our simulation results provide some evidence that sample selection influences econometric estimates of compensating wage differentials and supports Smith's and Viscusi's contention that the differential for workplace hazards is lower in samples weighted toward high-risk occupations.[7]

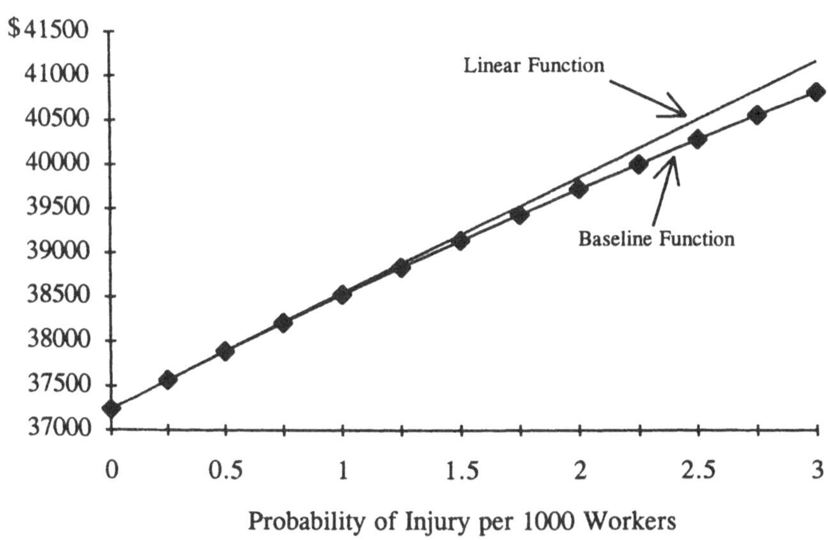

Figure 3.4 Baseline Hedonic Equilibrium Wage Function (1993 dollars)

Implied Value of Life

A key dimension of any governmental policy to improve workplace safety is determining whether the added benefits of fewer injuries outweigh the costs of additional investments in safety. To perform a cost-effectiveness calculation policy makers must place a value on human life. Economists suggest using a figure derived from the actual willingness of people to pay for additional safety. In our context workers willingly sacrifice higher pay for higher safety. In particular, Table 3–2 shows that workers in perfectly safe establishments sacrifice $1,315 to avoid a 1 in 1,000 chance of a workplace injury resulting in death. Put slightly differently, 1,000 workers would collectively pay $1,315,000 to avoid the possibility one would die, creating an implied value of life of $1,315,000 (1993 dollars).

Estimates of the value of life obtained from observed compensating wage differences for death risk in the workplace vary widely. A recent review of the empirical literature (Viscusi 1993) found the highest value of life estimate exceeded the smallest estimate by a factor of more than twenty ($17.9 million versus $0.7 million, 1993 dollars). Viscusi (1983) argued that the variation in estimated values workers implicitly place on their lives reflects mostly differences among studies in the risk measurements used and mix of workers included. The remaining variability would mirror the concavity of the hedonic wage function. With a concave wage function studies using samples composed of persons accepting high risks of injury would find low values of life and studies using samples composed of persons requiring low risks of injury would find high values of life. Dillingham (1985) contended that if risk were measured precisely researchers would find the true range of value of life estimates to be about 2 to 1.

The calculations in Table 3–2 show a steady decrease in the value of life as safety falls but the range is far from 2 to 1. One possible explanation might be that residual mismeasurement of risk continues to exaggerate the range of empirical values. In other words, our results may give a truer picture of the diversity of value of life estimates in the economy. Alternatively, another set of parameter values may produce a hedonic function that more closely corresponds to Dillingham's 2 to 1 range or even the 20 to 1 range noted by Viscusi.

SENSITIVITY CHECKS

To examine the importance of our initial assumptions we decreased the value of each parameter by 1 percent, holding all other parameters constant, and calculated the percentage change in the wage and the compensation for risk at each level of workplace safety.[8] Table 3-3 presents the results of our sensitivity checking exercise as impact elasticities calculated from the continuous version of the hedonic model.

Risk Aversion and Risk Levels

Based on the numbers presented in Table 3-3 the most critical assumptions underlying the computed hedonic equilibrium wage function are those concerning the distribution of workers. The elasticities of wages and compensation for risk with respect to the risk aversion parameter are several orders of magnitude larger than the elasticities of wages and compensation for risk with respect to any of the other parameters. In addition to the distribution of workers only the distribution of firms, the average injury frequency, and the maximum level of risk appear important in explaining the initial values calculated for risk compensation. However, most of the elasticities in Table 3-3 are less than 0.50 in absolute value. Radically increasing either the absolute level of wages or the compensation for risk by altering the four underlying parameters appears impossible.

Input Substitutability and Heterogeneity Distributions

As another check on our initial results we also examined the quantitative importance of two other background assumptions: (i) that the

Table 3-3. Sensitivity Checks: Wage and Compensation for Risk Elasticities

Parameter	Elasticity	$\pi =$ 0	$\pi =$ 0.00025	$\pi =$ 0.0005	$\pi =$ 0.00075	$\pi =$ 0.001	$\pi =$ 0.00125	$\pi =$ 0.0015	$\pi =$ 0.00175	$\pi =$ 0.002	$\pi =$ 0.00225	$\pi =$ 0.0025	$\pi =$ 0.00275	$\pi =$ 0.003
α_{max}	Wage	0	0.147	0.281	0.402	0.511	0.607	0.691	0.763	0.822	0.870	0.905	0.928	0.939
	Comp	16.768	15.803	14.790	13.725	12.604	11.422	10.174	8.855	7.456	5.970	4.389	2.706	
α_{min}	Wage	0	0.107	0.216	0.328	0.442	0.558	0.676	0.795	0.915	1.036	1.156	1.277	1.397
	Comp	12.198	12.892	13.593	14.300	15.012	15.725	16.438	17.148	17.850	18.541	19.214	19.863	
$(\alpha_{max} - \alpha_{min})$	Wage	0	0	0	0	0	0	0	0	0	0	0	-0.001	-0.001
	Comp	0.006	0.004	0.001	-0.001	-0.004	-0.006	-0.009	-0.012	-0.014	-0.017	-0.020	-0.023	
y (starting at y = $1,000)	Wage	0	0.001	0.002	0.003	0.003	0.004	0.005	0.006	0.006	0.007	0.008	0.008	0.009
	Comp	0.090	0.090	0.090	0.091	0.091	0.091	0.091	0.091	0.091	0.091	0.091	0.091	
μ_{max}	Wage	0	-0.004	-0.008	-0.010	-0.011	-0.011	-0.011	-0.011	-0.009	-0.008	-0.007	-0.006	-0.005
	Comp	-0.497	-0.380	-0.269	-0.164	-0.068	0.017	0.087	0.141	0.173	0.178	0.152	0.090	
$M(0)$	Wage	0	0.005	0.008	0.010	0.012	0.012	0.012	0.011	0.010	0.009	0.007	0.006	0.005
	Comp	0.518	0.398	0.283	0.175	0.075	-0.015	-0.090	-0.148	-0.183	-0.191	-0.165	-0.100	
$M(\pi^m)$	Wage	0	0	0	-0.001	-0.001	-0.001	-.001	-0.001	-0.001	-0.001	0	0	0
	Comp	-0.032	-0.025	-0.019	-0.012	-0.006	-0.001	0.005	0.009	0.012	0.013	0.012	0.009	
Elasticity of Substitution	Wage	0	0	0	0.001	0.001	0.002	0.002	0.002	0.002	0.002	0.002	0.001	0.001
	Comp	0.017	0.036	0.049	0.056	0.057	0.049	0.034	0.010	-0.020	-0.053	-0.080	-0.079	
Average Frequency of Injury	Wage	0	0	0.001	0.001	0.003	0.005	0.007	0.009	0.011	0.014	0.016	0.017	0.018
	Comp	-0.013	0.059	0.126	0.186	0.237	0.278	0.306	0.319	0.312	0.282	0.224	0.135	
π^m	Wage	0	0.009	0.017	0.024	0.031	0.037	0.042	0.048	0.053	0.058	0.063	0.068	0.074
	Comp	0.003	-0.058	-0.114	-0.164	-0.206	-0.238	-0.257	-0.262	-0.247	-0.210	-0.145	-0.049	

Note: Table 3-1 lists the functions and parameter values used in the simulations. Except for the parameter under investigation, all parameters were set at initial values. Elasticities were calculated using 1 percent decreases in the initial values. To determine the nonlabor income elasticities, we examined a 1 percent decrease in y from $1,000 ($1,000 represents approximately 10 percent of a worker's wage income). The wage and compensation for risk elasticities for ν, δ, γ, total number of workers, and total number of establishments were less than 0.001 and 0.035, respectively, at all injury probabilities.

$\pi \equiv$ the probability of a work-related injury resulting in total disability or death.

Comp $\equiv$ the slope of the wage function evaluated for a 1/1000 increase in the probability of work-related injury.

partial elasticity of substitution is the same between all input pairs and (ii) that interfirm and interperson heterogeneity are each described by uniform distributions of parameters α and μ. Concerning the elasticity of substitution it may be the case that safety equipment is a relatively fixed factor and is therefore less substitutable for the other two inputs. Halving the partial elasticities of substitution σ_{sn} and σ_{sk} reduces wages along the hedonic locus by at most 0.06 percent. With regard to work and firm heterogeneity we reformulated the model using normal distributions for α and μ, holding their mean values at the original levels while examining several different values for the second moments. Again, the changes in the computed hedonic wage function were minor when we changed the shape of the parameter distributions describing worker and firm heterogeneity.

Long-Run Utility and Profit Maximization.

As a final test of the robustness of our initial results we examined a major structural change in the numerical model. Specifically, to increase realism we assumed workers attempt to maximize lifetime expected utility and firms attempt to maximize the present value of long-run expected profits. Following the discussion in the previous paragraph we also modeled worker and firm heterogeneity using normal distributions for α and μ and the production of output using a nested constant partial elasticity of substitution production function. We provide a complete description of the model and the associated parameter values in Appendixes 4A and 4B.[9] To move the results further away from the baseline situation examined previously we also calibrated the model to produce a 1 in 10,000 average frequency of injury, a rate approximating that found in most studies using industry (versus occupational) fatality data (Viscusi 1993).

Figure 3.4 presents the baseline wage function with long-run utility maximizing workers and profit-maximizing firms. The most striking feature of the new equilibrium is the relatively smaller rise in wages but larger drop in risk compensation than in our baseline simulations in Table 3-2. As the probability of injury grows from 0 to 3 in 10,000,

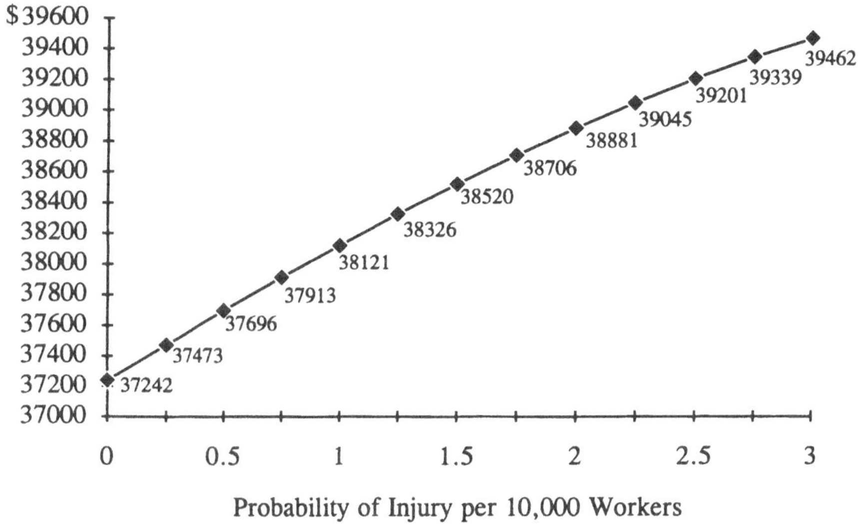

Figure 3.5 Baseline Hedonic Equilibrium Wage Function Calibrated to Produce a 1 in 10,000 Average Frequency of Injury (1993 dollars)

wages increase from $37,242 to $39,462 (1993 dollars) and risk compensation decreases from $924 to $492. Because we now evaluate risk compensation using a 1 in 10,000 increase in the probability of injury the implied value of life ranges from $4.9 to $9.2 million. Most empirical studies have found a much larger value of life than the $1,315,000 we calculated in the original formulation of the numerical model. The results in Table 3-4 are much closer to the majority of existing studies. Additionally, the range of values from low to high risk also quite closely corresponds to Dillingham's projected 2 to 1 ratio. In short, in a context devoid of mismeasurement we observe the same likely range as Dillingham believes would be produced econometrically across samples if a consistent measure of fatal injury risk were used.

As in the original formulation of the numerical model, to determine the sensitivity of the results we decreased each parameter value in the numerical model by a small percent, holding all other parameters constant, and recalculated the wage function. The resulting elasticities essentially duplicate those presented in Table 3-3. Although the revised model allowing long-run utility and profit maximization generates a much larger value of life, the initial equilibrium is no more sensitive to changes in the underlying parameter values than the original formulation. Hence, both versions of the numerical model should generate roughly the same qualitative and quantitative conclusions about public policies geared toward improving workplace safety.

CALIBRATING THE SIMULATION MODEL

In the remainder of the book we calibrate the simulation models using occupational fatality rates instead of industry fatality rates. By using the most extreme value for the probability of a workplace fatality we may underestimate the value of life but overestimate the frequency of injury. In both cases the bias may be relatively small, however. Our numerical simulation model attempts to capture the situation facing manufacturing firms in the late 1960s. Based on Kniesner and Leeth (1991b), the rate of workplace fatalities (primarily accidents) in manufacturing is higher than the average for all industries (4 in 10,000 versus 1 in 10,000) and the value of life is lower ($0.7 million versus $2 to $17 million, 1993 dollars). Furthermore, the 1 in 1,000 probability of a workplace death we use is not far from the rate Gegax, Gerking, and Schulze (1991) found in their survey of worker's perceived fatality risk on the job (9 in 10,000). Considering that industrial data unfortunately miss most workplace diseases, using a 1 in 1,000 rate of workplace fatalities for our numerical simulations seems most realistic.

CONCLUSION

We have discussed the crucial details involved in research that numerically simulates hedonic labor market equilibrium and disturbances to that equilibrium. When workplace safety levels are continuous hedonic labor market equilibrium can be described by the simultaneous solution to three first-order differential equations with associated boundary conditions. The first simulation method we considered numerically solved the complete system of differential equations. Alternatively, when job safety levels are discrete equilibrium is described by a set of simultaneous nonlinear supply and demand equations. The second method we examined numerically solves the supply–demand system.

Although the discrete and continuous simulation techniques have quite different theoretical underpinnings we found the two simulation methods produce a similar picture of the relationship between wages and workplace safety across firms. Both modeling approaches showed wages rising by almost 10 percent as the probability of a work-related injury expands from 0 to 3 in 1,000. Likewise, all results demonstrated a general reduction in risk compensation as the level of safety falls, but that the risk compensation reduction is relatively minor. Based on the slope of the market wage function, a worker accepting a 1 in 1,000 increase in risk earns from $1,054 to $1,315 more per year, a gain of about 3 percent.

Researchers who attempt numerical simulations of the type we present should note that assumptions concerning the exogenous distribution of workers seem to have the most critical influence on the numerical results. Calibrating the model to varying injury rate scenarios may also produce dramatic shifts in the calculated value of life. A 1 in 1,000 chance of a workplace fatality (the average for high-risk occupations) produced about a $1 million value of life, while a 1 in 10,000 chance (the average across all industries) produced about a $7 million value of life. Regardless of which injury rate scenario one chooses the dynamics of the model appear relatively unaffected. The specific computational technique used to determine the equilibrium sorting of workers and firms among safety levels and to calculate equilibrium wages also appears to be relatively unimportant. In the remainder of the book we use the continuous version of the numerical model calibrated to produce a 1 in 1,000 probability of injury to evaluate workplace safety policies. Alternative specifications of the numerical model should alter the quantitative results only slightly and not change the qualitative conclusions.

APPENDIX 3A: MATHEMATICAL DERIVATION OF THE CONTINUOUS MODEL

From Rosen (1974), we know labor market equilibrium occurs within a given risk interval when

$$L(\alpha)\left|\frac{d\alpha}{d\pi}\right|d\pi = N(\mu)F(\mu)\left|\frac{d\mu}{d\pi}\right|d\pi, \qquad \textbf{(A3.1)}$$

where:
$L(\cdot)$ ≡ the density function of workers with respect to α, with $\alpha_{min} \leq \alpha \leq \alpha_{max}$,
α ≡ a utility function parameter determining an individual's aversion to risk,
π ≡ the probability of a work-related death or total disability,
$N(\cdot)$ ≡ the demand for labor by an individual firm,
$F(\cdot)$ ≡ the density function of firms with respect to μ, with $\mu_{min} \leq \mu \leq \mu_{max}$, and
μ ≡ a parameter representing the efficiency of safety equipment in the production of output.

Using the first-order conditions we can demonstrate that both $d\alpha/d\pi$ and $d\mu/d\pi$ are negative. Thus, we can convert equation (A3.1) into the following expression describing the sorting of workers into levels of injury risk

$$\frac{d\alpha}{d\pi} = \frac{N(\mu)F(\mu)}{L(\alpha)}\frac{d\mu}{d\pi}. \qquad \textbf{(A3.2)}$$

To solve equation (A3.2) we need expressions for $N(\mu)$ and $d\mu/d\pi$. By totally differentiation the three first-order conditions for expected profit maximization (equations 2.4–2.6) with respect to π, treating μ as an endogenous variable, we can derive an expression for the equilibrium sorting of firms into the various levels of risk. The sorting rule is

$$\frac{d\mu}{d\pi} = \left\{ -c_0 \frac{\partial^2 R}{\partial n^2}\frac{\partial^2 R}{\partial k^2} - 2c_1 \frac{\partial^2 R}{\partial k\,\partial n}\frac{\partial^2 R}{\partial k\,\partial \pi} + c_0\left(\frac{\partial^2 R}{\partial k\,\partial n}\right)^2 + c_1^2 \frac{\partial^2 R}{\partial k^2} + \frac{\partial^2 R}{\partial n^2}\left(\frac{\partial^2 R}{\partial \pi\,\partial k}\right)^2 \right\} \Big/ c_2, \tag{A3.3}$$

where

$$c_0 \equiv \frac{\partial^2 R}{\partial \pi^2} - \left(\frac{d^2 W}{d\pi^2}(1-\pi) - 2\frac{dW}{d\pi}\right)n - p_s\frac{d^2 S}{d\pi^2} - \frac{d^2 P_b}{d\pi^2}bn - \frac{d^2 V}{d\pi^2},$$

$$c_1 \equiv \frac{\partial^2 R}{\partial \pi \partial n} - \frac{dW}{d\pi}(1-\pi) + w - \frac{dP_b}{d\pi}b,$$

$$c_2 \equiv \frac{\partial^2 R}{\partial \pi\,\partial \mu}\frac{\partial^2 R}{\partial n^2}\frac{\partial^2 R}{\partial k^2} + c_1\frac{\partial^2 R}{\partial k\,\partial n}\frac{\partial^2 R}{\partial k\,\partial \mu} + \frac{\partial^2 R}{\partial \pi\,\partial k}\frac{\partial^2 R}{\partial k\,\partial n}\frac{\partial^2 R}{\partial n\,\partial \mu} - \frac{\partial^2 R}{\partial \pi\,\partial \mu}\left(\frac{\partial^2 R}{\partial k\,\partial n}\right)^2 - c_1\frac{\partial^2 R}{\partial n \partial \mu}\frac{\partial^2 R}{\partial k^2} - \frac{\partial^2 R}{\partial \pi\,\partial k}\frac{\partial^2 R}{\partial n^2}\frac{\partial^2 R}{\partial k\,\partial \mu},$$ and

$R(\cdot)$ ≡ the expected revenue function,
n ≡ the number of workers,
k ≡ the quantity of capital,
s ≡ the quantity of safety equipment; $s = S(\pi)$, $S' < 0$,
p_b ≡ the per-unit price of WC insurance; $p_b = P_b(\pi)$, $P_b' > 0$,
b ≡ WC insurance benefits,
w ≡ $W(\pi)$; $W(\cdot)$ is the hedonic wage function, and
v ≡ $V(\pi)$; v represents the expected level of fines for violating government safety and health standards.

The equilibrium values of n and k are jointly determined by the first-order conditions for expected profit maximization. Treating both μ

and π as exogenous, we can represent the optimal usage of labor and capital with the following general functions

$$n = N(\mu, \pi, w, p_k) \text{ and} \quad \textbf{(A3.4)}$$

$$k = K(\mu, \pi, w, p_k). \quad \textbf{(A3.5)}$$

The final two expressions required to close the model are the first and second derivatives of the hedonic wage function. By rearranging the first-order condition for expected utility maximization (equation 2.2) we can show that

$$\frac{dW}{d\pi} = \frac{U(\cdot) - \tilde{U}(\cdot)}{(1-\pi)\dfrac{\partial U}{\partial w}}, \quad \textbf{(A3.6)}$$

where $U(\cdot)$ ≡ the worker's indirect utility function if uninjured, $u = U(w + y;\alpha)$,

$\tilde{U}(\cdot)$ ≡ the worker's indirect utility function if dead or totally disabled; $\tilde{u} = \tilde{U}(b + y;\alpha)$, $U > \tilde{U}$, and $U' > \tilde{U}'$, and

y ≡ nonlabor income.

We can also establish the required second derivative by totally differentiating (A3.6) with respect to π, treating α as endogenous. This second derivative is

$$\frac{d^2W}{d\pi^2} = \left\{\frac{\partial U}{\partial \alpha}\frac{d\alpha}{d\pi} - \frac{\partial \tilde{U}}{\partial \alpha}\frac{d\alpha}{d\pi} - (1-\pi)\frac{\partial^2 U}{\partial w \partial \alpha}\frac{d\alpha}{d\pi}\frac{dW}{d\pi} + 2\frac{\partial U}{\partial w}\frac{dw}{d\pi} - (1-\pi)\frac{\partial^2 U}{\partial w^2}\left(\frac{dW}{d\pi}\right)^2\right\} \Big/ (1-\pi)\frac{\partial U}{\partial w}. \quad \textbf{(A3.7)}$$

After substituting equations (A3.4), (A3.5), and (A3.7) into (A3.3) and equation (A3.4) into (A3.2), the system of equations describing hedonic equilibrium becomes

$$\frac{d\alpha}{d\pi} = A'(\pi, \alpha, \mu, w, M'), \tag{A3.8}$$

$$\frac{d\mu}{d\pi} = M'(\pi, \alpha, \mu, w, A', W'), \text{ and} \tag{A3.9}$$

$$\frac{dW}{d\pi} = W'(\pi, \alpha, w). \tag{A3.10}$$

As with any system of differential equations boundary conditions are necessary to determine a unique solution. The assumptions made concerning the expected profit function, $G(\pi, n, k; \mu)$, imply that the productivity parameter, μ, increases both profit and the optimal level of safety. Thus, the zero profit constraint endogenously determines the minimum productivity of safety equipment in the production of goods (μ^m) and, through the simultaneous solution to the first-order conditions, the maximum level of risk (π^m). Firms with $\mu < \mu^m$ would choose levels of risk greater than π^m but cannot earn even zero economic profits and, therefore, do not operate. Few firms exist at any given level of π, except with some bunching at $\pi = 0$; competition prohibits any massing of workers at π^m, which would drive wages to zero. Finally, the existence of economic profit at $\pi = 0$ permits massing of firms and workers at the zero risk level. The wage rate at $\pi = 0$ equilibrates the supply and demand for workers. In the model we develop boundary conditions establish labor market equilibrium at the zero risk and maximum risk levels.

Mathematically, the boundary conditions are

$$\begin{aligned} R(\pi^m, n^m, k^m; \mu^m) - W(\pi^m)(1-\pi^m)n^m - p_k k^m \\ - p_s S(\pi^m) - P_b(\pi^m) b n^m - V(\pi^m) - c = 0, \end{aligned} \tag{A3.11}$$

$$\frac{\partial R}{\partial \pi}(\pi^m, n^m, k^m; \mu^m) - \frac{dW}{d\pi}(\pi^m)(1-\pi^m)n^m + W(\pi^m)n^m$$
$$- p_s \frac{dS}{d\pi}(\pi^m) - \frac{dP_b}{d\pi}(\pi^m)bn^m - \frac{dV}{d\pi}(\pi^m) = 0, \quad \textbf{(A3.12)}$$

$$\frac{\partial R}{\partial \pi}(\pi^m, n^m, k^m; \mu^m) - W(\pi^m)(1-\pi^m) - P_b(\pi^m)b = 0, \quad \textbf{(A3.13)}$$

$$\frac{\partial R}{\partial k}(\pi^m, n^m, k^m; \mu^m) - p_k = 0, \quad \textbf{(A3.14)}$$

$$A(\pi^m) - \alpha_{\min} = 0, \text{ and} \quad \textbf{(A3.15)}$$

$$\int_{A(0)}^{\alpha_{\max}} L(\alpha)d\alpha = \int_{M(0)}^{\mu_{\max}} N(\mu)F(\mu)d\mu, \quad \textbf{(A3.16)}$$

where n^m and k^m represent (A3.4) and (A3.5) evaluated at π^m.

Equations (A3.11)-(A3.15) define the maximum probability of injury, π^m, and the accompanying equilibrium, while equation (A3.16) establishes equilibrium at the zero level of injury risk. Viewed independently the six equations determine the labor market outcomes of interest, which are $M(\pi^m)$, π^m, n^m, k^m, $A(\pi^m)$, and $W(0)$.

APPENDIX 3B: STRUCTURAL EQUATIONS AND INITIAL PARAMETER VALUES

Having presented in Appendix 3A the technical details of how the labor market establishes hedonic equilibrium we now fill in the details of the structural equations underlying the matching of workers to jobs by injury risk. We fill out the structure of our numerical simulations by describing the quantitative properties of workers' utility functions including the distribution of workers by risk preferences, firms' profit functions including the distribution of employers by their costs of injury reduction, and the economic properties of product and input markets.

Workers

We represent a worker's expected utility as

$$\bar{u} = (1-\pi)(w+y)^{1-\alpha} + \pi\beta(b+y)^{1-\alpha}, \qquad \textbf{(B3.1)}$$

where π is the probability of a work-related injury, α and β are utility function parameters, and w, y, and b are wage income, nonlabor income, and workers' compensation insurance (WC) benefits.

Worker heterogeneity is completely captured by interpersonal differences in the value of α. Increases in α increase the Arrow-Pratt measures of absolute and relative risk aversion. Because differences in the value of α meter differences in attitude among individuals regarding job safety the power utility function in (B3.1) is a convenient form. Risk aversion requires α be distributed from 0 to 1. In Thaler and Rosen (1975) the marginal worker is indifferent to employment with a zero incremental probability of death and employment with a 0.001 probability of death if compensated with an additional \$176 per year (1967 dollars). Based on a \$176 compensation for risk and the expected utility function (B3.1) the marginal worker has an α of 0.963. To capture inframarginal workers we used a uniform distribution for α ranging from 0.96 to 0.98. Workers with high values of α sort into safer jobs whereas workers with low values of α sort into more hazardous employment.

The incentive to self protect is less dependent on the attitude toward risk than the incentive to insure (Ehrlich and Becker 1972). Although α is positively related to traditional measures of risk aversion high α workers may sort into high-risk versus low-risk jobs. A high α not only decreases the marginal value of hazard pay causing a worker to choose safer employment, but also decreases the utility loss from an injury giving a worker the incentive to accept less safety. On balance, the dominant effect depends on the entire set of parameters used in the numerical model. In every case we examined, the results indicated a positive sorting between α and workplace safety ($1-\pi$).

In the expected utility function (B3.1) β determines the loss of utility from a work-related accident or disease resulting in death or total disability, income held constant. Higher values of β imply greater ex ante utility from the provision for survivors, for example. Because the bequest motive is likely to be small we set $\beta = 0.1$ in the simulations. Workers experience a nine-tenths utility loss from an injury holding income constant.

As explained later, we define 1 unit of an input as the amount capable of generating $1 of net return to the firm. Because an individual worker earns more than $1 per year, we must define the number of efficiency units of labor each worker supplies. The average payroll expense in manufacturing was $9,000 per worker in 1972. Combined with our other assumptions we created a per worker payroll of approximately the 1972 U.S. manufacturing average by assuming each worker provides an employer with 10,000 efficiency units of labor. Thus, the hedonic wage function we calculated should be viewed as reflecting both the rental rate and the quantity of human capital per worker. Because nonlabor income comprises only a small fraction of a typical worker's earnings, we set y equal to zero in the initial simulations. We also set insurance benefits equal to zero initially. Although we assumed that workers in an industry are similar in terms of (i) human capital accumulation, (ii) nonlabor income, and (iii) insurance coverage in subsequent simulations we examined the labor market effects of both nonlabor income and WC insurance.

Firms

We used a three factor constant partial elasticity of substitution (CPES) production function to depict a firm's output. Specifically,

$$q = a\left[\delta(\mu s^{-\rho}) + \gamma k^{-\rho} + (1-\delta-\gamma)n^{-\rho}\right]^{-\upsilon/\rho}, \qquad \textbf{(B3.2)}$$

where
$a \equiv$ the efficiency parameter,
$s \equiv$ the quantity of safety equipment; $s = S(\pi)$, where $S(\cdot)$ is the safety production function,
$n \equiv$ labor,
$k \equiv$ capital,
$\delta, \gamma \equiv$ distribution parameters,
$\upsilon \equiv$ the returns to scale parameter,
$\rho \equiv$ the elasticity of substitution parameter, $\sigma = 1/(1+\rho)$, and
$\mu \equiv$ the parameter capturing heterogeneous technology, which influences the efficiency of safety equipment in the firm's production of output.

In general, a firm's profit function can be expressed as

$$\begin{aligned}\bar{g} = R(p_o, S(\pi), n, k; \mu) - W(\pi)(1-\pi)n - P_b(\pi)bn \\ - p_s S(\pi) - p_k k - V(\pi) - c,\end{aligned} \qquad \textbf{(B3.3)}$$

where
p_o, p_s, p_k, and $P_b(\pi) \equiv$ the per-unit prices of output, safety equipment, capital, and insurance; $P_b(\pi)$ is the insurance price function with $P_b{}' > 0$,
$V(\pi) \equiv$ the expected fine function for violating safety and health standards, $V' > 0$, and
$c \equiv$ total fixed costs.

Except for its explicit recognition of interfirm heterogeneity an equation such as (B3.3) is the starting point for all studies of the economics of industrial safety. Our research has firms differing only with respect to their ability to transform safety inputs into output for sale. Our numerical simulation model follows Rosen (1986) who notes that it is natural to think of the firm as engaging in joint production where it simultaneously sells a conventional good to the market and a nonmarket good to its workers. Although we follow Rosen's lead by emphasizing differences in output technology among firms our explicit formulation of the problem comes from Viscusi (1979, p. 120) who simply treats safety as an input into the production process with the usual positive and diminishing marginal product. Thaler and Rosen (1975) also model the situation in a generally similar manner. However, neither Viscusi nor Thaler and Rosen address the issue of whether differences in output technology or safety technology across firms are what produce a hedonic labor market equilibrium, and we are unaware of any econometric research clearly establishing the underlying reason(s) for interfirm safety heterogeneity.

We chose to follow Rosen (1986) and emphasize the output productivity aspect of firms' expenditures on safety. Thus, the crucial aspect of our research is that firms differ with respect to some attribute causing a significant variation in the net cost of safety. Firms with an output technology where safety equipment is highly productive can produce a given level of safety more cheaply than firms where safety equipment is less productive. The parameter μ determines the size of the output effect of safety equipment. Firm heterogeneity comes from assuming μ is uniformly distributed from 0 to 1. We assume firms have identical values for all other expected profit function parameters. Because the safety input measured in efficiency units increases with μ, a firm's purchase of safety equipment and its resulting level of workplace safety both increase with μ.

We standardized all prices in the simulation model to 1.0 except for the price of WC insurance. Similar to previous research in applied general equilibrium (Harberger, 1962; Shoven and Whalley 1972), a unit of capital, labor, or safety equipment is the amount generating \$1 of net return to the firm. We selected values for the production function's distribution parameters, δ and γ, roughly corresponding to the proportions of revenues generated by each of the three factors in the manufacturing sector of the United States. Although we do not know its exact proportion we assumed that the firm's income generated by safety equipment is relatively small. In the initial simulation we set the cost share parameters for safety equipment and capital (δ and γ) equal to 0.10 and 0.20.

Cross-section data for the U.S. manufacturing sector suggest 1.0 as a reasonable starting value for the elasticity of substitution of capital and labor (Hamermesh 1993). We also investigated how our results change with the elasticity of substitution, or as we move away from the Cobb-Douglas production function.

The returns to scale parameter limits the absolute size of firms when the price of output is constant. We selected υ such that the elasticity of demand for labor by an individual firm is approximately unitary.

For internal consistency we chose a value for the efficiency parameter (a) creating a \$1 rental rate on human capital in completely safe workplaces. Because workers are assumed to have identical quantities of human capital, differences in wages among firms then come from differences in hazard pay rather than differences in innate worker productivity.

We also had to specify the safety production function to solve for the equilibrium hedonic wage function numerically. As explained earlier we determined the three coefficients of the quadratic safety production function by calibrating the model to produce an initial equilibrium with an average incremental probability of total disability of 1/1000, a maximum probability of total disability of 3/1000, and a proportion of firms without incremental workplace hazards of 0.40.

To complete our numerical description of firms' expected profits we needed to specify the fixed costs of production (c). Because we used a decreasing returns-to-scale CPES production function, computed profits are extremely large. We reduced profits so the last firm in the labor market earns only a normal rate of return by adding a payment representing fixed costs. By setting the payment to a value high enough to keep some potential firms out of the market we could examine the impact of public policy programs on firms' entry and exit behavior. The value selected for fixed costs implies a value of μ of 0.1 for the marginal employer. We denote the value of μ for the marginal firm by $M(\pi^m)$. The range of μ observed initially in the labor market based on expected profits and fixed production costs is 0.1 to 1.0.

Input and Output Markets

We assumed both input and output markets are competitive. Firms purchase capital and safety equipment at fixed prices of $1. We implicitly assumed both nonlabor inputs are produced in constant cost industries, implying their prices are invariant to the particular changes in the economic environment we consider in the numerical simulations.

We described the output market with a unitary elastic demand function. We initially set the price of output at $1. Later, when we investigated the labor market effects of public policies geared toward improving industrial safety we allowed the price of output to adjust. Hence, changes in labor costs have direct effects and indirect effects on hedonic labor market equilibrium through their impact on product price.

Finally, our numerical simulation used a fixed number of workers and establishments. Because labor is measured in terms of dollars of net return to the firm, we used payroll expenditures in the manufacturing sector of the United States divided by the efficiency units of labor per worker to represent total employment. We then set the number of firms equal to the number of manufacturing establishments. To reproduce the situation studied by Thaler and Rosen we based both the numbers of workers and firms on 1970 data from the *Statistical Abstract of the United States*.

NOTES

[1]The conclusion requires $\left[d\left(\overline{U}_i - \overline{U}_{i+1}\right)\right]/d\alpha > 0$, where $\overline{U}_i \equiv$ expected utility in submarket i. To be consistent with the definition of risk aversion we

assumed that α strictly increases the difference in expected utility between higher and lower levels of workplace safety.

[2]The sign of $d(G_i - G_{i+1})/d\mu$ is equal to the sign of $-d^2G/d\pi\, d\mu$. By totally differentiating the indirect profit function we see that

$$\frac{d^2G}{d\pi\, d\mu} = -\left(\frac{\partial^2 R}{\partial n \partial \mu}\frac{dN}{ds} + \frac{\partial^2 R}{\partial k \partial \mu}\frac{dK}{ds} + \frac{\partial^2 R}{\partial s \partial \mu}\right)\frac{dS}{d\pi},$$

where K is the reduced-form equation for capital and other functions have been defined previously. Further, by totally differentiating the first-order conditions for profit maximization we can demonstrate that dN/ds and dK/ds are strictly positive when the mixed partial derivatives of the revenue function are positive. Thus, assuming μ increases the marginal revenue product of all inputs and $dS/d\pi < 0$ the difference in expected profit between higher and lower safety levels is strictly increasing in μ.

[3]Labor demand equals zero if the firm cannot earn at least zero economic profit.

[4]To reduce computation time we relied on the firm's first-order conditions for profit maximization to solve for μ at the j injury rates and then used the calculated differentials to approximate $d\mu/d\pi$ (M') in the remaining equations of the model. As we showed in Kniesner and Leeth (1988), using the first-order conditions to solve equation (3.3) versus using the formal mathematical characterization of the differential equation alters none of the simulation results.

[5]Figure 3.3 draws heavily from Figure 5.1 in Shoven and Whalley (1992, p. 104).

[6]In the continuous model a boundary condition endogenously determines the maximum level of risk. However, in the discrete version the maximum level of risk is exogenous, so the assumption $\pi^m = 0.003$ is insufficient to determine the safety production function coefficients uniquely. Certain values for the safety production function coefficients cause some of the submarkets to be noneconomic (demand zero labor) We used the requirements of a 1/1000 average frequency of injury and a 40 percent incidence of firms offering zero incremental risk to determine s_0 and s_1 in the discrete model. We then varied s_2 to identify the range of values where production occurs in all submarkets and then set s_2 equal to the midpoint of its range.

[7]Empirical research concerning the shape of the hedonic wave function is limited. The one study directly examining the wage function's shape found the function to be concave in injury risk. Further, the percentage decrease in the compensation for risk from safe to relatively hazardous jobs is approximately the same as we calculated (Leigh and Folsom 1984).

[8]We varied the calibration factors to satisfy the initial conditions. For instance, in examining the impact of the maximum level of risk aversion we varied the safety production function coefficients, the output efficiency parameter, product demand price elasticity, and the level of fixed costs so that the average injury rate was 0.001, the maximum level of risk was 0.003, the wage rate for completely safe jobs was $10,000, the price of output was $1, the proportion of completely safe workplaces was 0.40, and the value of μ for the least profitable firm in the market was 0.1.

[9]For simplicity we typically used uniform distributions to represent worker and firm diversity. To allow a more thorough check of the sensitivity of our results we also simulated the case where worker risk preferences have a truncated normal distribution with a mean of 0.6, a standard deviation of 0.1, and minimum and maximum values of 0.4 and 0.8. Likewise, we assumed the efficiency of safety in the production of output has a truncated normal distribution with a mean of 0.5, a standard deviation of 0.25, and minimum and maximum values of 0 and 1.

4 THE SAFETY POLICY SYSTEM

The issue of workplace safety in the United States is crucially linked to two public policies intended to reduce worker injury and disease incidence: (1) state workers' compensation insurance (WC) and (2) the federal Occupational Safety and Health Act of 1970 (OSHA). In general, economists have studied OSHA and WC separately.[1] Feldstein (1977) and others, however, have argued that researchers need to devote more attention to examining and describing public policy interactions given the potential for subtle and not so subtle counterbalancing effects. Using the numerical simulation framework described in the previous chapter, we now examine how mandatory insurance (WC) and the direct regulation of transaction characteristics (OSHA) interact in the labor market to affect long-term equilibrium outcomes with a view toward identifying straightforward safety enhancing refinements in the safety policy package.

We found OSHA standards significantly reduce industrial injuries only when quite stringently enforced. For instance, reducing workplace injuries by as much as 14 percent requires regulators impose fines for safety and health violations more than 40 times greater than the fines originally imposed in the early 1970s. We found little interaction between OSHA and WC. Most importantly, we discovered that the degree of experience rating of WC insurance premiums crucially influences the impact of WC benefits on workplace safety. With incomplete experience rating WC actually increases slightly the number of serious work-related injuries, while with complete experience rating WC greatly reduces the number of serious work-related injuries. In fact, when insurance companies fully experience rate WC insurance premiums the safety enhancing effect of WC exceeds that of OSHA standards.

NUMERICAL SIMULATION DETAILS

A numerical solution of hedonic labor market equilibrium requires specific functional forms to represent workers' preferences, firms' outputs, and the distribution functions of risk aversion and efficiency of safety measures. Appendix 4A fully describes the numerical model. Table 4-1 provides a summary listing of the model's structural equations and initial parameter values. In choosing starting values for the parameters underlying the model we reproduced labor market equilibrium just prior to the enactment of OSHA.

The model we used to study the interface between OSHA and WC differs from the model developed in Chapter 3 in two respects. First, to increase realism we assumed workers attempt to maximize lifetime expected utility and firms attempt to maximize the present value of long-run expected profits. Second, to study the impact of possible public policy changes we incorporated specific relationships representing OSHA activities and the WC insurance system.

OSHA Enforcement Activities

We represent OSHA enforcement activities by the expected level of fines for safety and health violations. To examine OSHA's impact on wages and workplace injuries we combined historical information on fines with assumptions concerning the specific functional relationship between fines (ν) and injury probabilities (π).[2] Similar to the process used to generate the safety production function in Chapter 3 we begin by assuming a quadratic expected fine function, $\nu = \nu_0 + \nu_1\pi + \nu_2\pi^2$. For purposes of numerical simulation, we had to determine the intercept of the expected fine function, ν_0, and the coefficients of risk and its square, ν_1 and ν_2.

The intercept in the fine function captures the expected fine for a firm with no incremental safety and health hazards. If OSHA accurately assesses workplace hazards it will not cite perfectly safe firms for violating safety and health standards, and ν_0 should equal 0. The coefficient of the square of risk measures the degree to which OSHA targets hazardous work sites for additional inspections or additional fines for violations. If ν_2 equals 0 then a 1 percent increase in a firm's probability of injury increases its expected fine by 1 percent, whereas if ν_2 exceeds 0 than a 1 percent increase in a firm's probability of injury increases its expected fine by more than 1 percent. A unitary elasticity implies OSHA either inspects hazardous work sites more frequently or discovers a greater number of violations at each work site, and a larger elasticity implies even greater targeting efforts. In our initial simulations we assumed a unitary fine/risk elasticity and set ν_2 equal to 0. We also investigated how targeting influences both the labor market effects of OSHA and its interface with WC.

The final coefficient we had to determine is ν_1. In 1973, 98.7 percent of OSHA fines were for nonserious violations of standards, 1.2 percent were for serious violations, and 0.1 percent were for willful violations. On average, OSHA inspectors wrote 6.8 citations per firm and assessed penalties of $52, $2,109, and $3,593 (1993 dollars) for nonserious, serious, and willful violations. In short, OSHA penalized the typical *inspected* firm $553 (1993 dollars). Even a fine of $553 considerably overstates expected penalties because most firms faced only a very slight probability of an inspection. In OSHA's early years the largest manufacturing plants, ones with over 1,000 employees, stood less than a 10 percent probability of inspection. With a 10 percent chance of seeing an OSHA inspector and an average fine of $553 if inspected the typical firm in 1973 faced an expected penalty of only $55 for violating safety and health standards. In the simulations to

Table 4-1. Base Case Functions and Parameter Values

I. FUNCTIONS

Expected utility:

$$\bar{u} = \sum_{j=1}^{t}\left(\frac{1}{1+r}\right)^{j-1}\left\{(1-\pi)^j w^{1-\alpha} + \left(1-(1-\pi)^j\right)\left[\psi\beta(wi)^{1-\alpha} + (1-\psi)\beta y_a^{1-\alpha}\right]\right\}$$

$$+ \sum_{j=t+1}^{d}\left(\frac{1}{1+r}\right)^{j-1}\left\{\begin{array}{l}(1-\pi)^e\left[(1-\pi)^{t-e} y_s^{1-\alpha} + \left(1-(1-\pi)^{t-e}\right)\beta y_s^{1-\alpha}\right] \\ + \left(1-(1-\pi)^e\right)\beta\tilde{y}_s^{1-\alpha}\end{array}\right\}$$

Production of output:

$$q = a\left[\delta(\mu s)^{-\rho} + \gamma k^{-\rho} + (1-\delta-\gamma)n^{-\rho}\right]^{-\upsilon/\rho}$$

Production of safety:

$$s = s_0 + s_1\pi + s_2\pi^2$$

II. WORKERS

Variable:	d	e	i	r	t	y_a	y_s	$\tilde{y}_s$
Value:	31	20	0.5	0.03	24	\$1,000	\$4,500	\$3,000

Variable:	α	β	ψ	Efficiency units of labor per worker
Value:	[0.4, 0.8]	0.1	0.1	10,000

DEFINITIONS

a	≡	the efficiency parameter
c	≡	fixed costs
d	≡	remaining life expectancy
e	≡	the additional number of years a worker must be employed to receive maximum social security benefits
i	≡	the proportion of income replaced through WC insurance,
k	≡	capital
$M(0)$	≡	the value of μ for the last firm locating at $\pi = 0$
$M(\pi^m)$	≡	the value of μ for the firm locating at the maximum level of injury risk
n	≡	labor
p_b	≡	the price of WC insurance per unit of coverage
p_k	≡	the price per unit of capital
p_o	≡	the price of output
p_s	≡	the price per unit of safety equipment
q	≡	output
r	≡	the subjective rate of time preference and the real rate of interest
s	≡	safety equipment
t	≡	the number of years before retirement

III. FIRMS

Variable:	a	c	s_0	s_1	s_2
Value:	545	$\$6.65 \times 10^6$	3.50×10^5	-2.61×10^7	1.81×10^9

Variable:	v	γ	δ	μ	υ	ρ
Value:	0	0.2	0.1	[0, 1]	0.1	0.01

IV. MARKETS

Variable:	p_k	p_s	η	p_b	Number of workers	Number of establishments
Value:	\$1	\$1	-1	2.29×10^{-3}	1.69×10^7	356,344

V. DISTRIBUTIONS

Workers are distributed uniformly with respect to α using the density function:

$$L(\alpha) = 4.234 \times 10^8 \,|\, 0.4 \le \alpha \le 0.8, \quad L(\alpha) = 0 \text{ elsewhere.}$$

Firms are distributed uniformly with respect to μ using the density function:

$$F(\mu) = 3.56344 \times 10^5 \,|\, 0 \le u \le 1, \qquad F(\mu) = 0 \text{ elsewhere.}$$

VI. INITIAL CONDITIONS

W(0) = \$10,000 (\$37,242, 1993 dollars) $p_0 = \$1$ $\pi^m = 0.003$

Average injury rate = 0.001 $M(0) = 0.64$ $M(\pi^m) = 0.1$

$\bar{u}$	$\equiv$	expected utility
v	$\equiv$	the expected fine for safety and health violations
w	$\equiv$	the wage rate
$W(0)$	$\equiv$	the wage rate in perfectly safe establishments
y_a	$\equiv$	the income received from public or private assistance if denied WC benefits
y_s	$\equiv$	maximum social security benefits
$\tilde{y}_s$	$\equiv$	minimum social security benefits
α	$\equiv$	the degree of risk aversion parameter
β	$\equiv$	the fraction of pre-injury utility, income held constant
γ,δ	$\equiv$	distribution parameters
η	$\equiv$	the product price elasticity
μ	$\equiv$	the efficiency of safety equipment in the production of output
υ	$\equiv$	the returns to scale parameter
π	$\equiv$	the probability of a work-related injury resulting in total disability or death
π^m	$\equiv$	the maximum level of injury risk
ρ	$\equiv$	the elasticity of substitution parameter, $\sigma = 1/(1+\rho)$

represent OSHA's initial impact we varied the value of v_1 to produce an average fine of $553 for inspected firms or an expected fine of $55 for all firms. Using an average fine equal to the actual value in 1973 allows us to compare our results to existing econometric research to discover how well the numerical model replicates what we already know about OSHA.

In addition to an average fine of $553 we investigated the effects of average OSHA fines of $1,107, $3,980, and $22,329 (1993 dollars). In each case we varied v_1 to produce the actual average penalty for inspected firms.[3] Practically speaking, the numerical exercise simulates a variety of OSHA activities such as additional inspections, greater violations detected per inspection, and larger fines per violation. Each of the activities increases a firm's expected fine for violating safety and health standards, which encourages greater safety activities. Because the last three levels of possible OSHA enforcement are quite far from the values appearing in cross section and time-series data for the United States we would be unwilling to accept econometric forecasts of such extreme changes in OSHA's parameters. (Remember the error associated with an econometric forecast grows with the distance of the independent variables from their sample mean values.) Hence, the numerical simulation results we present sharpen the empirical picture we have of how OSHA affects labor market outcomes. Additionally, the simulation model allows us to investigate the important question of why OSHA has failed to noticeably improve workplace safety. Is OSHA ineffective because of only lax enforcement of safety and health standards or is OSHA ineffective because of a low economy-wide response rate to safety incentives?

Workers' Compensation Insurance

WC insurance influences workplace safety in two conflicting ways. Higher benefits for income loss if injured reduce workers' monetary concerns with injury leading them to accept higher risk jobs. In the other direction, higher benefits raise firms' costs of workplace injuries, leading them to provide greater safety. The more closely insurance companies match WC premiums to workplace safety the greater are firms' financial incentives to provide safer work environments. To model the labor market impact of WC we had to specify the level of benefits workers receive if injured and then detail the relationship between workplace safety and insurance premiums.

The large dispersion in the income replacement rate across type of injury, level of income, and state complicates specifying a single level of WC benefits. We based our initial value on Johnson and Wai's (1980, p. 38) estimate of the median replacement rate for married men (30 percent) as the most representative for the U.S. manufacturing sector in the early 1970s. Because Johnson and Wai ignored taxes in their study we used an income replacement rate of 0.50 in our baseline simulations to reflect federal and state income taxes of 40 percent.

We used three additional income replacement rates in investigating the labor market effects of WC benefits — 80, 100, and 150 percent. The 80 percent replacement rate reflects the recommendation of the National Commission on State Workmen's Compensation Laws. The two higher rates represent complete protection and substantial overprotection against income loss from a work-related injury. By using two higher rates of income replacement we could investigate much larger changes in WC benefits than possible from econometric estimates with existing data. Moreover, in our research the causality clearly runs *from* WC benefit changes *to* changes in workplace safety. We avoid possible simultaneous equations bias that may appear in econometric estimates when researchers are unable to separate two phenomena: (1) the influence of interstate differences in WC generosity on interstate differences in work-related injuries, and (2) the influence of interstate differences in workplace injury rates on WC generosity across states through the political system. By simulating the complete set of labor market adjustments to changes in WC our numerical results can provide an important addition to the quantitative picture of how WC insurance benefits affect workplace safety.

As we will discuss in greater detail in the next chapter WC insurance does not completely cover all work-related disabilities. A worker must demonstrate that an injury resulted from work activities to receive benefits. In many instances the actual cause of injury is elusive. Even if the worker realizes an injury is job related he or she may not be able to prove it. Because of imperfect state verification workers must anticipate the probability of receiving WC benefits when accepting employment in a hazardous workplace. In our simulations we adopted a 10 percent probability that an injured worker will receive WC benefits.[4]

In a competitive insurance market prices reflect the cost of benefits, the administrative expense of writing coverage, and the cost of health care. In our baseline simulation we assumed that all firms were charged the same premium per unit of coverage, so that experience rating is absent. The price adjusts so that the per year revenue collected by the insurance industry just covers the discounted expected costs of coverage. We assumed that the administrative and health care costs are a constant 50 percent of benefit payments, and the real rate of interest is 3 percent. Later in the chapter we discuss the impact of experience rating insurance premiums.

NUMERICAL SIMULATION RESULTS

Figure 4.1 depicts the baseline hedonic wage function, and Figure 4.2 shows the distribution of workers across the risk spectrum. Wages range from \$37,242 (1993 dollars) per year in perfectly safe establishments to \$40,791 per year in the most hazardous establishments. Based on the wage function, workers require a 3.6 percent premium to move from a zero risk establishment to an establishment offering the mean level of risk ($\pi=0.001$), which is well within the range of empirical estimates of compensating wage differentials (Smith 1979).

Hedonic labor market equilibrium can have a massing of workers and firms only at the zero incremental risk level. As seen in Figure 4.2, 41 percent of the workforce chooses employment in firms that are no more dangerous than everyday life. In our baseline hedonic labor market equilibrium 55 percent of workers choose what could be labeled low-risk jobs ($0 \leq \pi \leq 0.001$), 19 percent choose what we will call moderate-risk jobs ($0 < \pi \leq 0.002$), and 26 percent choose high-risk jobs ($\pi > 0.002$).

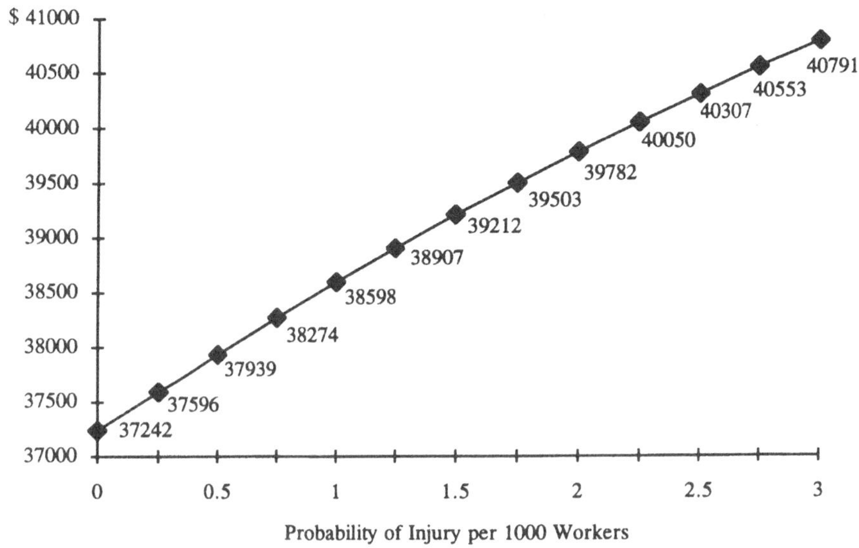

Figure 4.1 Baseline Hedonic Equilibrium Wage Function (1993 dollars)

The Labor Market Effects of OSHA

To investigate OSHA's labor market effects we increased the average fine for violating safety standards from 0 to (1) \$553, (2) \$1,107, (3) \$3,980, or (4) \$22,329. To examine the interaction between OSHA and the WC system we calculated OSHA's labor market effects at four levels of WC benefits (income replacement rates of 0.5, 0.8, 1.0, and 1.5).

Comparing the ceteris paribus impact of OSHA enforcement activities across levels of WC benefits allows us to determine the second derivative of each endogenous variable with respect to the two public policy programs. A large second derivative implies a substantial interaction effect between programs and argues for administering WC and OSHA as a single coordinated public policy package. In contrast, if

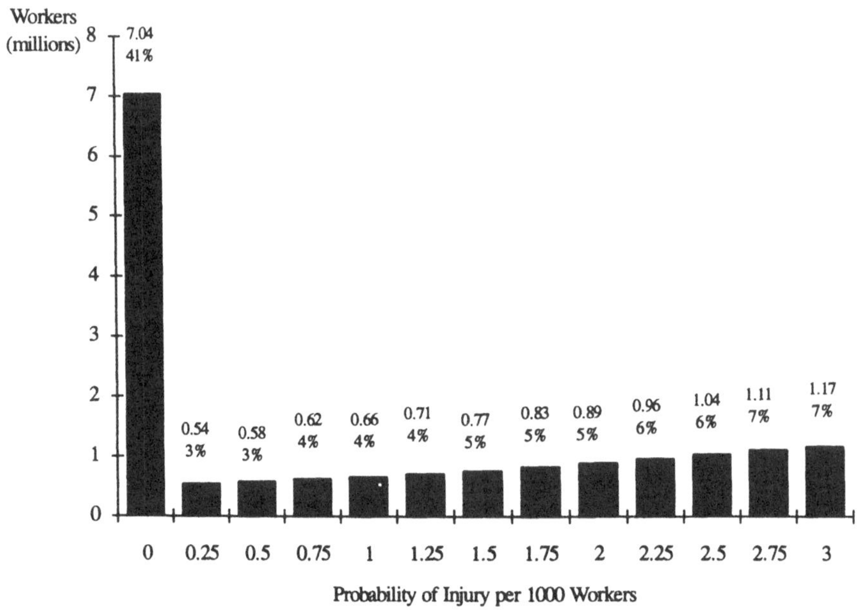

Figure 4.2 Baseline Employment Distribution

the interaction effects are small then no apparent public policy purpose is served by coordinating the two programs.

Job Safety. Table 4-2 indicates OSHA increases workplace safety, but the impact is large only with extreme levels of enforcement. Specifically, when WC insurance replaces 50 percent of income imposing an OSHA program with an average fine of $553 (an expected fine of $55) reduces the frequency of serious injury in the workplace by a mere 0.3 percent; fines of $1,107 and $3,980 reduce the average frequency of injury by only 0.6 and 2.3 percent. Only the largest average fine produces a significant improvement in workplace safety. Imposing a $22,329 fine for violating safety and health standards reduces the frequency of injury by just over 14 percent.

Econometric research has discovered at best weak relationships among measures of OSHA activities and industrial injury rates (See Chapter 2). Based on our calculations OSHA's impotence is not sur-

Table 4-2. Simulation Results: The Effects of OSHA Enforcement Activities on the Probability of Work-Related Injuries[a, b]

Average Fine for OSHA Safety and Health Violations	Rate of Income Replacement under WC	Average Frequency of Injury	Largest Probability of Injury
$553	0.5	-0.0031	-0.00565
		(-0.31)	(-0.19)
	0.8	-0.0031	-0.0056
		(-0.31)	(-0.18)
	1.0	-0.0031	-0.0055
		(-0.31)	(-0.18)
	1.5	-0.0031	-0.0055
		(-0.31)	(-0.18)
$1,107	0.5	-0.0062	-0.0112
		(-0.62)	(-0.37)
	0.8	-0.0062	-0.0112
		(-0.62)	(-0.37)
	1.0	-0.0062	-0.0111
		(-0.62)	(-0.37)
	1.5	-0.0062	-0.0111
		(-0.61)	(-0.37)
$3,980	0.5	-0.0228	-0.0410
		(-2.28)	(-1.37)
	0.8	-0.0227	-0.0408
		(-2.26)	(-1.36)
	1.0	-0.0227	-0.0407
		(-2.25)	(-1.35)
	1.5	-0.0226	-0.0405
		(-2.23)	(-1.34)
$22,329	0.5	-0.1417	-0.2599
		(-14.17)	(-8.66)
	0.8	-0.1412	-0.2586
		(-14.06)	(-8.60)
	1.0	-0.1409	-0.2578
		(-14.00)	(-8.56)
	1.5	-0.1402	-0.2562
		(-13.86)	(-8.48)

[a]Absolute (percentage) changes from the absence of an industrial safety and health program.

[b]To Improve readability, the average frequency of injury and the largest probability of injury have been increased by 10^3. Table 4-1 lists the functions and parameter values used in the numerical simulations.

prising.[5] More importantly, even vigorously enforced safety and health standards do not dramatically reduce injuries. Apparently, to reduce serious accidents and diseases by 2 percent requires more than a 700 percent increase in fines from the initial OSHA level of $553. A 14 percent reduction requires more than a 4000 percent increase in fines. The fundamental problem with OSHA is the extremely low labor market response rate to changes in the economic incentives to expand safety. As we demonstrate later the low response rate occurs in part because OSHA enforcement activities alter the hedonic wage loci leading to an increase in employment at moderate levels of risk.

Table 4–2 reveals only small interaction effects between OSHA and WC in reducing workplace health hazards on average; a marginally enforced OSHA program lowers the average frequency of injury by about the same amount at all four WC benefit levels (about 0.3 percent). However, a more vigorously enforced OSHA program, such as one with an average fine of $22,329, reduces injuries by slightly more when WC benefits are low (14.2 percent with current benefits) than when benefits are high (13.9 percent with maximum benefits).[6] The implied second derivative of the average frequency of injury with respect to both large changes in OSHA enforcement activities (an increase in the average fine from 0 to $22,329) and large changes in WC benefits (an increase in the rate of income replacement from 0.5 to 1.5) is only 0.0015, which means that WC benefits reduce OSHA's effectiveness in improving safety by at most 1 percent.

The impact of OSHA enforcement activities on the largest probability of injury appears in the last column of Table 4–2. The increase in fines for safety and health violations drives some marginally profitable firms out of business and encourages others to reduce workplace hazards. Because the least profitable firms have the most hazardous workplaces, both the effects within and between firms eliminate the highest-risk jobs. Fines of $553 lower the largest probability of injury by 0.2 percent. Not surprisingly, higher average fines magnify the reduction in high-risk jobs. Fines of $22,329 reduce the largest probability of injury by about 9 percent. OSHA activities increase overall safety partly due to the relatively large drop in extreme workplace hazards. As we document later, in a labor market context the *sole* reason OSHA improves safety is through the elimination of high-risk jobs; other market responses to OSHA *increase* the frequency of work-related injuries.

Finally, we can see from Table 4–2 WC reduces only slightly OSHA's effectiveness in eliminating extreme workplace hazards. Fines of $22,329 lower the largest probability of injury by 8.7 percent when WC replaces half of a worker's pre-injury earnings and lower it by 8.5 percent when WC provides 1.5 times the workers' pre-injury earnings. In other words OSHA is only slightly more effective in reducing the most severe workplace hazards when WC benefits are low.

Wages. Figure 4.3 depicts how OSHA affects wages at the various levels of injury risk. To avoid repetition we present simulation results only for the most stringently enforced OSHA program. The other three levels of enforcement activities produce smaller, but qualitatively

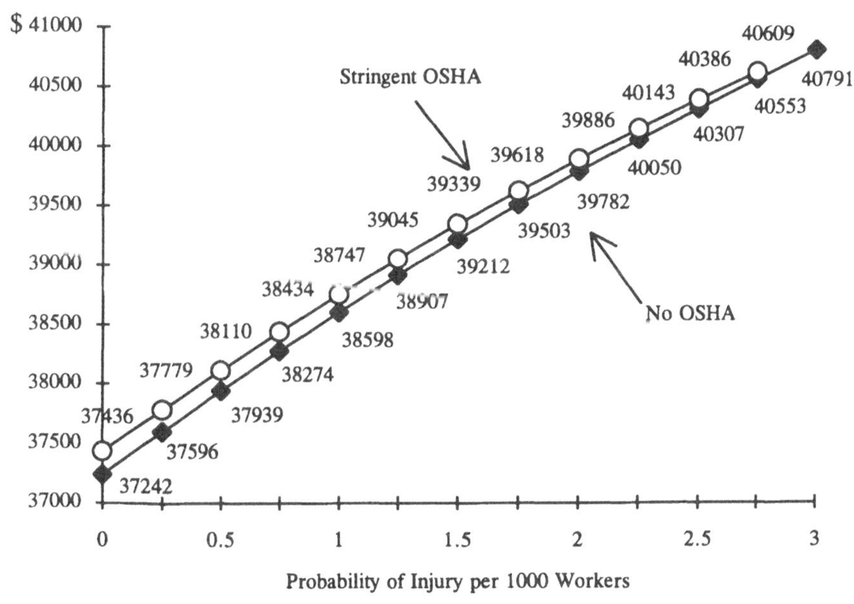

Figure 4.3 The Effect of Stringent OSHA Enforcement Activities on Wages (1993 dollars)

similar, effects. Because the interaction between OSHA and WC is minuscule where wage levels are concerned, we only show the wage function for the baseline level of WC benefits.

As can be seen in Figure 4.3 OSHA increases wages at each level of risk (shifts the equilibrium hedonic locus upward), with the largest increases coming at the lowest probabilities of injury. Specifically, at recent levels of WC benefits, fines of $22,329 increase wages by $194 per year (0.52 percent) in perfectly safe establishments and by $56 per year (0.14 percent) in the most hazardous establishments. Although it may appear counterintuitive for OSHA to increase wages, safety standards actually reduce the wage rates individual firms face. By installing more safety equipment firms decrease their probability of injury and move down the initial hedonic wage function. Employers react to the lower wages that now go with the higher levels of safety by increasing their desired labor input. With few firms going out of business because of higher expected costs and with aggregate labor supply fixed in the simulations, the rise in labor demand by firms installing greater safety equipment increases the overall demand for labor and raises wages at each risk level.[7]

Figure 4.3 also reveals that higher levels of OSHA enforcement flatten the hedonic wage function. By reducing the slope OSHA lowers the added income workers receive for accepting greater job hazards. Before OSHA a worker accepting an average risk job ($\pi=0.001$) received $1,356 more per year than a worker in a perfectly safe establishment. With stringently enforced safety and health standards the added compensation for risk falls 3 percent ($1,311). The reduction in risk compensation encourages workers to seek safer employment and so allows the movement of employers from high to low risk to be matched by a movement of workers. OSHA reduces risk compensation because it increases firms' costs of workplace hazards but OSHA leaves unchanged workers' willingness to accept hazards.

Employment. Figure 4.4 shows the impact of stringently enforced OSHA safety and health standards on the distribution of workers across safety levels at two levels of WC benefits, income replacement rates of 0.5 and 1.5. At both levels of WC benefits OSHA reduces the number of workers who locate in the uppermost segment of the risk distribution

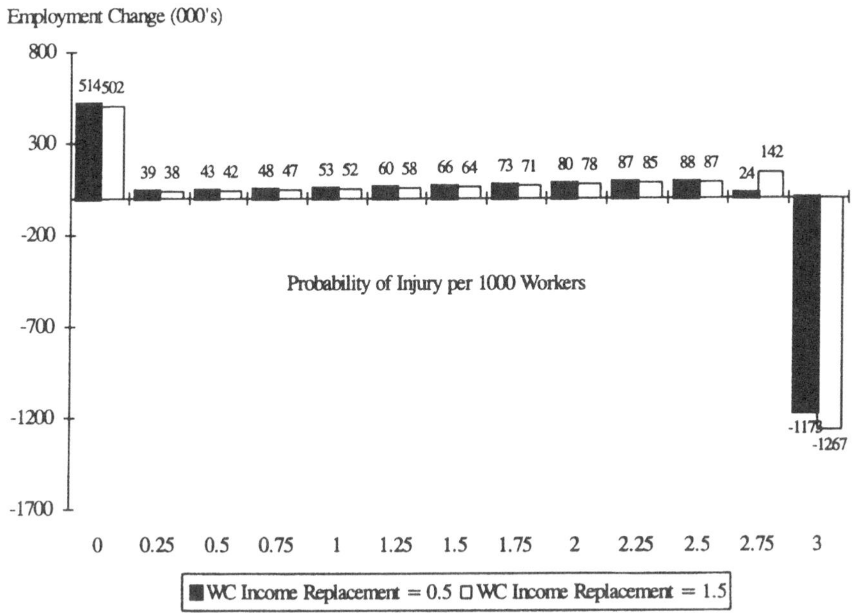

Figure 4.4 The Effect of Stringent OSHA Enforcement Activities on the Distribution of Employment

and increases the number of workers in medium- and low-risk jobs. To elaborate, OSHA decreases injuries in two ways. First, safety standards eliminate extremely high-risk jobs, as demonstrated by the reduction in the largest probability of injury in Table 4–2. Second, OSHA standards produce a sorting effect, which is a movement of workers from high-risk to medium- and low-risk jobs.

Table 4–3 separates the injury effects of the two forces.[8] Note the overall reduction in injuries occurs because the highest risk jobs are eliminated. The sorting effect increases the number of injuries because greater enforcement activities raise the number of workers on medium- and low-risk jobs. By focusing on a representative worker or firm, other theoretical and empirical research has ignored the sorting effect. When viewed in a *market* context, as we do here, OSHA eliminates

Table 4-3. The Effect of Stringent OSHA Enforcement Activities on the Number of Injuries Separated by Cause

Rate of Income Replacement	The Total Change in Injuries	The Change in Injuries Caused by Eliminating High-Risk Jobs	The Change in Injuries Caused by the Resorting of Workers
0.5	-2,400	-3,658	1,258
0.8	-2,391	-3,640	1,249
1.0	-2,386	-3,628	1,242
1.5	-2,374	-3,606	1,232

certain job hazards, but increases the number of workers exposed to other hazards.

Some researchers have argued on theoretical grounds that OSHA may lower the average level of safety because workers reduce personal safety efforts in response to increased firm efforts (Viscusi 1979, Rea 1981). The sorting effect reported in Table 4–3 does not rely on workers taking fewer precautions. In our simulations no worker accepts a higher level of risk on his or her new job and no firm becomes less safe. Instead, the sorting effect increases the number of work-related injuries because workers do not uniformly sort toward the lowest possible level of (incremental) risk, $\pi=0$. The approximately 1 million workers exposed to the greatest hazards prior to OSHA enforcement efforts in our model ($\pi > 0.00275$) move to the next highest level of risk ($0.0025 < \pi \leq 0.00275$) in response to marketwide changes in wages. Workers originally in the second-highest risk interval also move toward greater safety, but the net effect is an increase in employment at risk level $0.0025 \leq \pi \leq 0.00275$. Moreover, although every risk interval experiences a net increase in employment, the largest proportionate increase occurs at medium-risk jobs, $0.001 \leq \pi \leq 0.002$. What we call the sorting effect partially counteracts the impact on safety caused by eliminating the highest risk jobs. In short, OSHA is 50 percent less effective in reducing workplace injuries because OSHA leads to a movement of workers to moderate levels of risk, rather than a lock step movement toward complete safety.

The Interface Between OSHA and WC. By examining Figure 4.4 we can also gain insights into the interaction between OSHA and WC. Although the impact is small, WC reduces OSHA's movement of workers from high to low risk. At current levels of coverage stringent OSHA enforcement decreases employment in extremely hazardous jobs ($0.002 < \pi \leq \pi^m$) by 974,700 workers, or 5.8 percent of the total workforce in our model. With WC benefits of 1.5 times pre-injury earnings, stringent OSHA enforcement decreases employment in the highest risk sectors by 953,000, or 5.6 percent of the workforce. The slight reduction in the flow of workers away from extreme hazards happens because higher WC benefits lower the income loss from injury. To see why workers move away from the most hazardous jobs remember that OSHA increases firms' net cost of workplace hazards, inducing them to increase workplace safety. Because firms move to lower positions on the risk spectrum, equilibrium requires workers do so too. Changes in the hedonic wage function equilibrating the quantities of labor supplied and demanded, therefore, lead workers to move to greater safety. With generous WC benefits workers are less concerned with job safety and fewer relocate away from high-risk jobs, though. By discouraging the movement toward safer jobs WC reduces the favorable effect of OSHA on workplace safety.

The Labor Market Effects of WC

To investigate the impact of workers' compensation insurance we used a method identical to that described for investigating the impact of OSHA enforcement activities. Specifically, we first numerically simulated hedonic labor market equilibrium at four rates of income replacement (0.5, 0.8, 1.0, and 1.5) assuming no experience rating of WC premiums and no safety and health standards and then calculated changes caused by increasing WC benefits from the usual 0.50 rate of income replacement. To examine the interaction between OSHA and WC we varied the average fine for safety and health violations from $0 to $553, $1,107, $3,980, and $22,329 (1993 dollars). At each level of OSHA we then calculated the labor market effects of changing WC benefits from the usual replacement rate of 0.50. The labor market effects we calculated represent additional changes in labor market equilibrium caused by WC benefits that diverge from their historical level, including income replacement rates two and three times greater than have existed, on average.

Job Safety. At every level of safety and health regulation in Table 4–4 we see a positive association between WC benefits and the average frequency of injury. Raising WC benefits to the level suggested by the National Commission on State Workmen's Compensation Laws increases the average injury rate from 0.4 to 0.6 percent. Complete coverage increases injuries from 0.7 to 0.9 percent, whereas overcompensating income loss raises injuries from 1 to 1.5 percent. The results in Table 4–4, which are from a WC system with no experience rating, are consistent with the observed positive association between WC benefits and workplace injury rates across states (See Chapter 2).[9]

Theoretically, WC coverage reduces the monetary loss from injury, which encourages workers to accept less safe employment. With experience rating (which is absent here) WC also raises the cost of work hazards to firms, however, encouraging them to provide greater safety. In general, the sorting of workers toward higher risk and firms toward lower risk disequilibrates the labor market and changes the hedonic wage function which, in turn, alters workers' and firms' decisions regarding workplace safety. In the labor market equilibrium we simulated here there is only a movement of workers toward riskier employment.

Most importantly the numerical results we present emphasize the low labor market response rate to changes in the economic incentives to expand workplace safety. A 200 percent increase in the rate of income replacement (from 0.5 to 1.5) produces only about a 1 percent increase in injuries at current levels of OSHA enforcement. Clearly, our results do not argue for lowering WC benefit levels to encourage job safety.

To dig deeper into the labor market influence of the workers' compensation insurance system Table 4–4 also illustrates how WC influences the largest probability of injury. Because workers are more

Table 4-4. Simulation Results: The Effects of Workers' Compensation Benefits on the Probability of Work-Related Injury[a, b]

Rate of Income Replacement Under WC	Average Fine for Safety and Health Violations under OSHA	Average Frequency of Injury	Largest Probability of Injury
0.8	0	0.0043 (0.43)	0.0073 (0.24)
	$553	0.0043 (0.43)	0.0073 (0.24)
	$1,107	0.0043 (0.44)	0.0075 (0.25)
	$3,980	0.0044 (0.45)	0.0075 (0.25)
	$22,329	0.0048 (0.56)	0.0086 (0.32)
1.0	0	0.0067 (0.67)	0.0115 (0.38)
	$553	0.0067 (0.67)	0.0115 (0.38)
	$1,107	0.0067 (0.68)	0.0116 (0.39)
	$3,980	0.0068 (0.70)	0.0118 (0.40)
	$22,329	0.0075 (0.88)	0.0136 (0.50)
1.5	0	0.0117 (1.17)	0.0206 (0.69)
	$553	0.0117 (1.17)	0.0207 (0.69)
	$1,107	0.0117 (1.18)	0.0211 (0.69)
	$3,980	0.0119 (1.21)	0.0211 (0.71)
	$22,329	0.0131 (1.53)	0.0243 (0.89)

[a]Absolute (percentage) changes from a 0.5 rate of income replacement under WC.
[b]To improve readability, the average frequency of injury and the largest probability of injury have been increased by 10^3. Table 4-1 lists the functions and parameter values used in the numerical simulations.

willing to accept job hazards when WC is available, the maximum level of workplace hazards increases. For instance, when OSHA fines are zero complete WC coverage (100 percent replacement) raises the largest probability of injury by 0.4 percent compared to a 50 percent income replacement rate under WC. The only reason WC increases injuries in our simulations is through its impact on extremely hazardous jobs; higher WC benefits actually reduce the number of workplace injuries at moderate levels of injury risk.

The Interface Between WC and OSHA. We can use the results reported in Table 4–4 to examine further the interface between WC and OSHA. The absolute size of the interaction must, of course, be the same as reported earlier when we discussed the labor market effects of OSHA enforcement activities. Still, because WC increases injuries only slightly, the interaction between the two safety programs appear larger when put on a percentage basis. For instance, a movement toward a 0.80 rate of income replacement increases the average frequency of injuries by 0.4 percent in the absence of safety standards, but by 0.6 percent with maximum enforcement of safety standards. Further, the increase in WC benefits raises the maximum probability of injury by 0.2 percent without OSHA but by 0.3 percent with an extremely stringent OSHA program.

Wages. Figure 4.5 shows the reduction in wages caused by increasing WC benefits. We present only results for an increase from the historical 0.5 level to a 1.5 rate of income replacement; other benefit changes produce qualitatively similar conclusions. Because OSHA enforcement activities produce minor changes in the effect of workers' compensation insurance on wages we also only present the results assuming a zero average fine for safety and health violations. At all injury rates enforcement of OSHA safety standards lowers the reduction in annual wages due to WC by no more than $130 compared to the situation in which OSHA is totally absent.

The effect of WC on wages is quite small, less than 0.5 percent at every level of risk. The small WC effect is not surprising because, on average, insurance costs increase by less than $94 per worker to expand coverage to a 1.5 income replacement rate. Moreover, the cost increase is borne not only by workers but also by firms and consumers

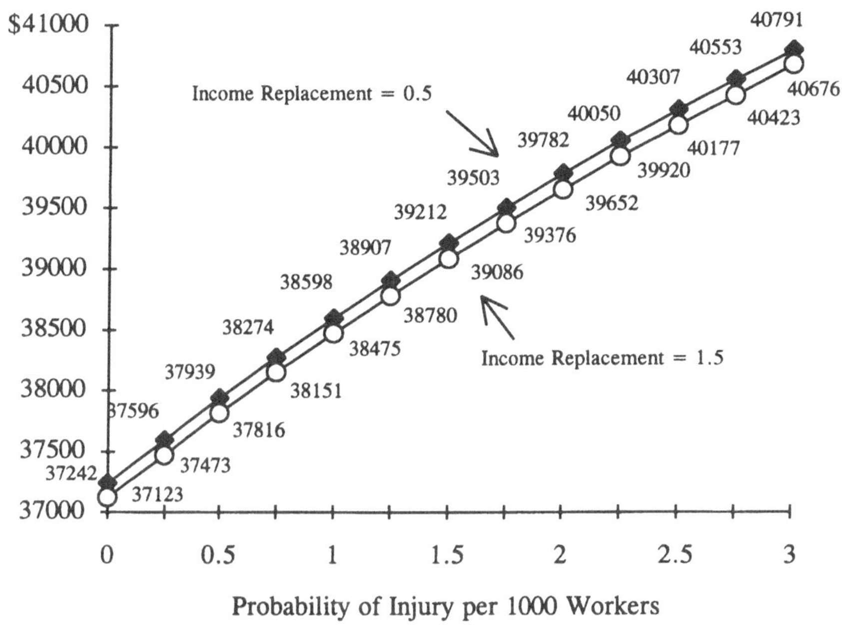

Figure 4.5 The Effect of Workers' Compensation on Wages (1993 dollars)

and, yet, our simulations imply workers bear a burden *exceeding* the increase in firms' insurance costs.[10] The difference between the workers' and firms' economic burdens reflects the net reduction in the purchase of safety equipment throughout the labor market. Because in our simulations safety equipment increases labor productivity the resulting decline in labor utilization further reduces wages.

Figure 4.5 also illustrates how higher WC benefits flatten the hedonic wage function, but the impact on the compensation for risk is tiny. Risk compensation must fall to reequilibrate the supply and demand for labor across the entire risk spectrum. Higher benefits for income loss encourages workers to accept greater job hazards but also encourages firms to provide safer workplaces. Only if risk compensation falls will the divergent interests of workers and firms coincide to reestablish labor market equilibrium. Based on the baseline function a worker accepting an average risk job receives $1,356 more per year than a worker accepting no job hazards. With a 1.5 rate of income replacement the compensation for risk falls to $1,352.

Employment. The final WC effect we consider concerns the distribution of workers across safety levels. Figure 4.6 shows the movement of workers across the risk spectrum caused by increasing the rate of income replacement from 0.5 to 1.5 with no enforcement and with maximum enforcement of OSHA safety and health standards. WC benefits lead workers to relocate away from relatively safe toward relatively hazardous workplaces. An increase in the rate of income replacement from its baseline 0.5 level to 1.5 attracts an additional 87,700 workers, almost 1 percent of the workforce in our model, to high-risk jobs ($0.002 < \pi \leq \pi^m$) when OSHA safety and health standards are totally absent. However, more generous WC benefits also

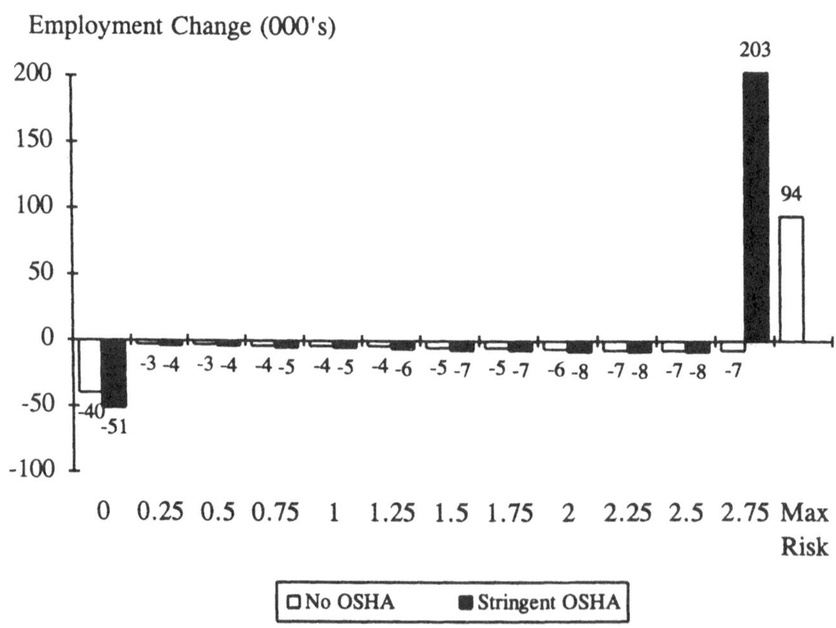

Figure 4.6 The Effect of Tripling Workers' Compensation Benefits on the Distribution of Employment

lower the number of workers employed at moderate levels of risk, in turn reducing the number of injuries within the moderate-risk segment of the risk spectrum.

Similar to OSHA, higher WC benefits affect injuries by (1) raising the maximum level of risk and (2) changing the sorting of workers among the various levels of safety. Because individual workers and firms do not expand their safety efforts, one might conclude the sorting effect reinforces the impact on extreme hazards. As Table 4-5 demonstrates, because of the sorting *away from* moderate levels of risk the number of injuries at the moderate risk levels falls with higher WC benefits, mitigating by more then 30 percent the increase caused by greater extreme hazards.

Table 4-5 also emphasizes the difference between examining a public policy question with a representative worker or firm and examining an entire market of heterogeneous workers and firms. Based on a representative worker or firm model higher WC benefits reduce safety incentives and, therefore, all categories of injuries should rise. However, as Figure 4.6 and Table 4-5 together show the movement of workers in the labor market among levels of safety results in a reduction of employment at moderate risk levels and, as a result, a reduction in injuries at moderate risk levels. Fewer injuries occur *not* because workers are more careful as firms reduce safety efforts, but because

Table 4-5. Simulation Results: The Effect of Workers' Compensation Insurance on the Number of Injuries Separated by Cause

Average Fine for Safety and Health Violations under OSHA ($)	The Total Change in Injuries	The Change in Injuries Caused by Eliminating High-Risk Jobs	The Change in Injuries Caused by the Resorting of Workers
$0	198	290	-92
553	198	290	-92
1,107	198	290	-92
3,980	202	297	-96
22,329	222	342	-120

more generous WC benefits produce less than a lock step movement by workers and firms toward greater workplace hazards.

Based on the distribution of workers we can also now better understand why OSHA regulations amplify the impact of WC insurance on injuries. As shown in Figure 4.6, with stringent enforcement of OSHA standards versus no enforcement, an additional 112,600 workers, an increase of about 150 percent, are attracted to the highest categories of risk ($0.002 < \pi \leq \pi^m$). OSHA enforcement efforts magnify the sorting of workers toward higher risk employment.

TWO POSSIBLE REFINEMENTS IN WC AND OSHA

Up to now we have seen that OSHA has only a scant safety enhancing effect and WC leads workers to accept somewhat more dangerous working conditions. We now explore the labor market effects of two possible program changes: (1) increasing the degree of experience rating of WC insurance premiums, and (2) increasing the emphasis OSHA places on inspecting high-risk establishments.[11]

Increasing the Experience Rating of WC Premiums

The National Commission on State Workmen's Compensation Laws recommended insurance providers experience rate WC premiums to the maximum extent feasible. Historically, WC insurers have not used the accident experience of a small establishment to determine the establishment's price of WC insurance. Instead, they have combined the injury statistics for all workers within an industrial class and determined a so-called manual rate for WC coverage. The product of each worker's wage rate and the manual rate for his or her industrial class summed over all workers determines the total WC insurance premium for a small firm. As firm size increases insurers pay less attention to the average experience of the industrial class (the manual rate) and more attention to the safety record of the individual firm. Experience rating of insurance policies refers to the process of adjusting insurance prices to reflect individual versus group experience. Currently, insurance providers completely experience rate WC premiums for only the largest establishments.

Some suggest the lack of complete experience rating contributes to the positive relationship observed between injury rates and WC benefits (Ruser 1985). In our initial simulations we assumed insurance prices reflect only group experience. Even with group rating, high-risk firms still pay greater *total* WC insurance premiums than low-risk firms because workplace hazards increase wages, which, in turn, increases the cost of WC coverage. Higher rates of income replacement magnify the indirect influence of workplace hazards on firms' insurance costs.

To examine the impact of experience rating we recomputed hedonic equilibrium for the situation in which insurance prices completely reflect the expected frequency of injury within a given establishment. From Table 4-6 we see that complete experience rating produces major differences in the calculated labor market effects of WC benefits. Unlike in the base case of incomplete experience rating, when there is complete experience rating of insurance premiums more generous WC benefits *improve* workplace safety. Because experience rating expands the impact of WC benefits on firms' marginal cost of injuries the resorting of firms toward greater safety dominates, in equilibrium, the resorting of workers toward less safe jobs. Increasing WC benefits to the level recommended by the National Commission on State Workmen's Compensation Laws reduces injuries by 8 percent and overcompensating income loss decreases injuries by 26 percent. In comparison, marginally enforced OSHA safety and health standards reduce injuries by 0.3 percent and stringently enforced standards reduce injuries by 14 percent. With complete experience rating of WC insurance premiums the safety enhancing effect of WC greatly exceeds that of OSHA.

The degree of experience rating produces only a small change in the calculated impact of OSHA on injuries. Less than a 1 percent difference exists between the results assuming no experience rating and complete experience rating. When insurance prices completely adjust to reflect a firm's expected frequency of injuries the calculated interaction effect between OSHA and WC is altered, though. In all cases other than complete experience rating WC benefits reduce OSHA's encour-

Table 4-6. Sensitivity Checks: The Impact of Two Possible Refinements in OSHA and Workers' Compensation Insurance on the Average Frequency of Injury

Small Changes in OSHA Enforcement and Workers' Compensation Benefits

Situation	Initial Enforcement of OSHA Standards (WC Benefits Replace 50% of Income)	WC Benefits Replacing 80% of Income (Zero Enforcement of OSHA Standards)	Joint Effect of OSHA Enforcement at Initial Levels and WC Benefits at 80% of Income	Interaction Between OSHA and WC
Base case	-0.0031	0.0043	0.0012	0
Complete experience rating of WC premiums	-0.0031	-0.0803	-0.0836	-0.0002
Greater targeting by OSHA of high-risk firms — a 1.5 expected fine/probability of injury elasticity	-0.0038	0.0043	0.0005	0

Large Changes in OSHA Enforcement and Workers' Compensation Benefits

Situation	Maximum Enforcement of OSHA Standards (WC Benefits Replacing 50% of Income)	WC Benefits Replacing 150% of Income (Zero Enforcement of OSHA Standards)	Joint Effect of OSHA Enforcement at Maximum Levels and WC Benefits at 150% of Income	Interaction Between OSHA and WC
Base case	-0.1417	0.0117	-0.1285	0.0015
Complete experience rating of WC benefits	-0.1412	-0.2589	-0.4441	-0.0440
Greater targeting by OSHA of high-risk firms — a 1.5 expected fine/probability of injury elasticity	-0.1817	0.0117	-0.1684	0.0016

[a] In each instance, we recalibrated the model to produce the initial conditions described in Table 4–1. Injury effects have been increased by 10^3.

agement of safety, and OSHA standards increase WC's discouragement of safety. With complete experience rating the two programs complement each other in reducing injuries. Maximum enforcement of OSHA standards decreases injuries an additional 31 percent at a 150 percent income replacement rate compared to a 50 percent income replacement rate. Moreover, increasing WC benefits from 50 percent to 150 percent of income lowers injuries an additional 17 percent with maximum enforcement of OSHA compared to no enforcement of OSHA regulations.

Improved Targeting of High Risk Establishments

In the second half of 1972 OSHA began targeting high-risk establishments for a greater number of inspections. Instead of randomly selecting firms for inspections they started to focus on firms in industries with records of frequent work-related injuries. (See Viscusi 1986b, 1986c for a description.) With OSHA targeting high-risk workplaces a firm's level of safety influences not only the fine if inspected but also the probability of inspection. In our simulation model a greater frequency of injury directly increases the expected fine for violating OSHA's safety and health standards. Thus, an increase in the expected fine can reflect both a higher probability of inspection and a higher fine if inspected. When computing our baseline hedonic equilibrium we assumed that a one percent increase in the probability of injury increases by one percent a firm's expected fine for violating OSHA standards, an expected fine/probability of injury elasticity that may be unrealistically low.

To examine the labor market impact of increasing the expected fine/probability of injury elasticity we varied the coefficients v_1 and v_2 in the quadratic equation describing the relationship between expected OSHA violations and the probability of injury. Specifically, we raised the fine/probability of injury elasticity to 1.5. Because we still hold the average fine constant, higher elasticities represent greater targeting of hazardous workplaces for OSHA inspections. With greater targeting the initial average fine of $553 for OSHA violation reduces injuries by approximately 0.4 percent, which is 22 percent more than originally calculated; the maximum average fine of $22,329 decreases injuries by just over 18 percent which is 28 percent more than originally computed. The results in Table 4–6 demonstrate targeting magnifies the labor market impact of OSHA, but greater targeting of OSHA inspections to high-risk workplaces produces no differences in the calculated impact of WC benefits and produces only minor differences in the interaction effects between OSHA and WC.

CONCLUSION

Our primary objective in Chapter 4 has been to investigate numerically the links among labor market outcomes and the joint properties of the two programs designed to improve workplace safety in the United States: (1) states' workers' compensation insurance programs and (2) the federal government's workplace safety standards of the Occupational Safety and Health Act of 1970. Changes in OSHA and WC alter workers' and firms' safety decisions and affect the equilibrium relationship between wages and workplace safety. By viewing the problem in a hedonic equilibrium context the resulting numerical simulations enhance our knowledge of how OSHA and WC work. More specifically, we incorporated the compete set of feedback effects on workers and firms and examined clearly exogenous and large changes in the two safety programs, which is difficult to do within the context of an econometric model.

We found OSHA safety and health standards significantly reduce industrial accidents and diseases only when quite stringently enforced. For example, to reduce injuries by 14 percent requires approximately a 4000 percent increase in enforcement efforts from those initially observed in 1972. We also found with incomplete experience rating of insurance premiums, WC benefits actually *increase* slightly the number of serious accidents and diseases. Because greater insurance benefits reduce income loss workers accept higher risk jobs, increasing employment in the high-risk sectors and, in turn, increasing the largest probability of injury. Under higher WC benefits the reduction in the compensation for risk outweighs the economic incentive for firms to provide greater safety. Additionally, we discovered a significant sorting effect by workers and firms among the various levels of safety mitigating both OSHA's positive impact and WC's negative impact on safety.

The interaction effects between the two public policy programs are relatively minor. Higher WC benefits reduce slightly OSHA's beneficial impact on workplace injuries, while higher levels of OSHA enforcement amplify slightly the effect of WC benefits on workplace injuries.

The major finding to emerge from our simulations in Chapter 4 is that the degree of experience rating of workers' compensation premiums influences both the direction of the effect of WC benefits on workplace safety and the interface between the OSHA and WC programs. If insurance providers were to experience rate WC premiums perfectly then WC would dramatically reduce workplace injuries, and OSHA and WC would strongly complement each other in enhancing workplace safety.

APPENDIX 4A: STRUCTURAL EQUATIONS AND INITIAL PARAMETER VALUES

Here we provide more details of the structure of our simulation model. The discussion to follow describes our numerical specification of the three key components of the model, which are workers' utility functions and heterogeneity in attitudes towards risk, firms' profit functions and heterogeneity in ability to jointly produce output and workplace safety, and the characteristics of how people and firms interact in the labor and product markets.

Workers

In modeling how workers choose among risky jobs we ignored for simplicity variation in accident severity. Workers base their employment decisions on the probability of an injury resulting in total incapacity for all future time periods. We also assumed workers initially expect to remain with an employer until retirement or until an injury occurs. In addition, we ignored the labor/leisure decision by workers (See Heckman and Sedlacek 1990 for a discussion of the importance of modeling labor force nonparticipation as an option). Each employer offers identical hours of work, and workers accept employment in only one establishment. Finally, we assumed all firms purchase the legally mandated level of workers' compensation insurance, which is not augmented by other types of disability coverage. Thus, workers choose among jobs based on the level of risk and not on the level of insurance coverage.

The problem facing the representative worker is finding the level of risk (π) that maximizes lifetime expected utility subject to $\pi \geq 0$. We represented a worker's expected utility using the strongly separable power utility function

$$\bar{u} = \sum_{j=1}^{t}\left(\frac{1}{1+r}\right)^{j-1}\left\{(1-\pi)^j w^{1-\alpha} + \left(1-(1-\pi)^j\right)\left[\psi\beta(wi)^{1-\alpha} + (1-\psi)\beta y_a^{1-\alpha}\right]\right\}$$

$$+ \sum_{j=t+1}^{d}\left(\frac{1}{1+r}\right)^{j-1}\left\{\begin{aligned}&(1-\pi)^e\left[(1-\pi)^{t-e} y_s^{1-\alpha} + \left(1-(1-\pi)^{t-e}\right)\beta y_s^{1-\alpha}\right]\\&+\left(1-(1-\pi)^e\right)\beta\tilde{y}_s^{1-\alpha}\end{aligned}\right\}, \quad \textbf{(A4.1)}$$

where
- $t \equiv$ the number of years until retirement,
- $r \equiv$ the subjective rate of time preference and the real rate of interest,
- $w \equiv$ the wage rate,
- $\alpha \equiv$ the degree of risk aversion,
- $\psi \equiv$ the probability of receiving workers' compensation benefits if injured,
- $\beta \equiv$ the fraction of pre-injury utility, income held constant,
- $i \equiv$ the proportion of income replaced through workers' compensation insurance,
- $y_a \equiv$ the income received from public or private assistance if denied workers' compensation insurance,
- $d \equiv$ remaining life expectancy,
- $e \equiv$ the additional number of years a worker must be employed to receive maximum social security benefits,
- $y_s \equiv$ maximum social security benefits, and
- $\tilde{y}_s \equiv$ minimum social security benefits.

Expected utility is the sum of the worker's utility if uninjured and the worker's utility if injured, weighted by the appropriate probabilities. Because we investigated extremely serious injuries, we assumed injured workers never return to the labor force. Thus, $(1-\pi)^j$ represents the probability a worker will not be injured during the first j time periods and $\left(1-(1-\pi)^j\right)$ represents the probability the worker will be injured.

Prior to retirement the worker's level of utility, if uninjured, depends on earned income ($w = W(\pi)$). If injured, the worker's utility depends on whether or not WC benefits are received. Because it is sometime difficult to determine both health status and cause of injury, some workers with a work-related total disability will not receive WC benefits. We assumed workers apply for WC benefits at the beginning of the period in which they are injured and have the probability ψ of receiving benefits. If the worker receives WC benefits, then he or she receives them until retirement. Otherwise, the worker receives other forms of public or private assistance. In addition, we assumed WC benefits are a fraction (ι) of previous earnings, whereas other forms of assistance are unrelated to previous earnings.

Expected utility, post retirement, depends on social security benefits and health status. To simplify computations we used only two social security benefit levels. If a worker continues working an additional e years, he or she receives benefits of $\$y_s$ per year until death. Otherwise, the worker receives the lower level of benefits ($\$\tilde{y}_s$) each year until death. The term $(1-\pi)^e$ represents the probability of remaining uninjured the number of years necessary to receive maximum social security benefits. Because health status also affects the well-being of retired workers, we must account for the probability of injury after completely fulfilling the work requirements of the social security system. With t years until retirement, $(1-\pi)^{t-e}$ reflects the probability of safety during the interim period, and $\left(1-(1-\pi)^{t-e}\right)$ reflects the corresponding probability of injury.

We assumed differences in the value of α across workers completely capture worker heterogeneity. Specifically, workers are treated alike in terms of: (1) quantity of human capital, (2) retirement age, (3) life expectancy, (4) subjective rate of time preference, (5) WC coverage, (6) probability of receiving WC benefits, (7) entitlement to public or private income maintenance programs, and (8) social security coverage. Because the Arrow-Pratt measures of absolute and relative risk aversion are both positively related to α, differences in α reflect differences in attitude among individuals regarding job-injury risk. Risk aversion requires α be distributed from 0 to 1. Regression results in presented in Thaler and Rosen (1975), when evaluated at the mean wage and incremental risk of death, imply α in equation (A4.1) equals 0.60 for the marginal worker. To capture inframarginal workers we used a uniform distribution for α ranging from 0.40 to 0.80. Workers with high values of α sort into safer jobs whereas workers with low values of α sort into more hazardous employment.

In our simulations we investigated the most severe category of injuries, those representing total incapacity during all future time periods. To represent the large loss in utility from such a severe accident we set β equal to 0.1. Workers experience a 90 percent utility loss from an injury holding income constant.

Because we used Thaler's and Rosen's estimates to determine the distribution of α, we also based the number of years to retirement and life expectancy on their study. The average age in their sample, men in relatively hazardous occupations, is 41. According to the *U.S. Statistical Abstract* a 41-year-old white male had an expected life span of 31 years in 1972. Thus, we set d equal to 31 and, using 65 as the age of retirement, set t equal to 24. In addition, we used 3 percent, which is approximately the real interest rate in the United States over the last 50 years, as the rate of time preference (r) in the worker's utility function (A4.1).

Finally, we had to initialize four other expected utility function parameters: (1) the proportion of income replaced through WC, (2) the probability of receiving WC benefits if injured, (3) the income received from other public assistance programs if denied WC benefits, and (4) the benefits from social security. We have already described income replacement and the likelihood of receiving WC earlier in the chapter. Concerning other public assistance, in 1971 maximum annual welfare payments ranged from about \$700 in Mississippi to slightly more than \$4,000 in Massachusetts, New York, and New Jersey (Reynolds, 1974, p. 324). Considering the range in public assistance levels across states, the level of WC benefits, and wages in perfectly safe jobs, we chose a value of \$1,000 (\$3,724, 1993 dollars) to represent public assistance to workers denied WC benefits. Public assistance of \$1,000 for an uncovered injured worker represents more than a 90 percent drop in income from pre-injury earnings and is only about 20 percent of the average WC recipient's benefits. Concerning social security, in 1972 average social security benefits replace approximately 45 percent of after-tax earnings. Reduced benefits due, say, to early retirement replaced roughly 30 percent of after-tax earnings (Schieber 1982). We assumed a worker employed past the age of 60 (e=20) receives maximum benefits (y_S). Using $10,000 ($37,242, 1993 dollars) as the base salary figure, we set y_S equal to $4,500 ($16,759, 1993 dollars). For workers injured before reaching 60 we set social security benefits ($\tilde{y}_S$) equal to $3,000.

Firms

The economic problem confronting the employer is choosing the combination of physical capital, labor, and workplace safety that maximizes shareholder wealth. As in Chapter 3 we used a three factor CPES production function to depict a firm's output.

$$q = a\left[\delta(\mu s)^{-\rho} + \gamma k^{-\rho} + (1-\delta-\gamma)n^{-\rho}\right]^{-\upsilon/\rho}, \qquad \textbf{(A4.2)}$$

where $a \equiv$ the efficiency parameter,
$s \equiv$ the quantity of safety equipment; $s = S(\pi)$, where $S(\cdot)$ is the safety production function,
$n \equiv$ labor,
$k \equiv$ capital,
$\delta, \gamma \equiv$ distribution parameters,
$\upsilon \equiv$ the returns to scale parameter,
$\rho \equiv$ the elasticity of substitution parameter, and
$\mu \equiv$ the parameter capturing heterogeneous technology, which influences the efficiency of safety equipment in the firm's production of output.

Safety programs are jointly productive inputs; additional safety equipment increases output and decreases deaths due to industrial injuries. The parameter μ determines the magnitude of the output effect. Firm heterogeneity is captured by assuming μ is uniformly distributed from 0 to 1. We took all other expected profit function parameters to be identical across firms. It is important to recognize that a firm's safety input, measured in efficiency units, increases with μ, which means a firm's purchase of safety equipment and in turn its level of workplace safety, is directly related to μ.

The firm's total expected annual cost includes the expected wage bill, the expense of WC insurance premiums, the cost of safety equipment, the cost of capital equipment, expected fines for possibly violating OSHA standards, and any fixed costs of production. The expected wage bill is the product of the total number of workers hired, the wage rate, and the expected fraction of workers uninjured. The total expense of WC insurance is composed of the unit price, the number of units of coverage purchased per employee, and the number of workers employed. We assumed each firm purchases the legally mandated level of WC insurance, expressed as a proportion of a worker's income replaced in the event of an injury. By multiplying the number of units of safety equipment and capital by their respective unit prices, we developed expressions for the economic use (depreciation) of inputs. Finally, we assumed OSHA fines are an increasing function of the probability of injury, investors are risk neutral, workers' rate of time preference equals the real rate of interest (r), and the representative worker hired by the firm (each time period) works t years.

Combining the production function in (A4.2) with measures of annual expected costs generates the shareholder wealth relationship

$$\bar{g} = \sum_{j=1}^{\infty}\left(\frac{1}{1+r}\right)^{j-1}\{p_o q - W(\pi)(1+\pi)n - P_b(\pi,\psi,r)itW(\pi)n - p_s S(\pi) - p_k k - V(\pi) - c\}, \tag{A4.3}$$

where
$p_b \equiv$ the price of insurance expressed as a fraction of the dollar coverage purchased; $p_b = P_b(\pi,\psi,r)$,
$p_s \equiv$ the price of safety equipment,
$p_k \equiv$ the price of capital,
$V(\pi) \equiv$ expected fines for safety and health violations, and
$c \equiv$ fixed costs.

Maximizing shareholder wealth is equivalent to maximizing the discounted sum of each year's expected economic profit. For simplicity we assumed that expected revenue and all cost items are time invariant, which means that examining a single year's expected profit function yields the optimal use of safety equipment, capital, and labor for all periods.

Similar to our approach in Chapter 3 we standardized all prices in the simulation model to 1.0, except for the price of WC insurance. We also set the production function's distribution parameters (δ and γ) equal to 0.10 and 0.20 and the elasticity of substitution ($\sigma = 1/(1+\rho)$) equal to 1.0. We adjusted the value of the production efficiency parameter (a) to create a $1 rental rate on human capital in completely safe establishments and the value of the returns to scale parameter (υ) to make the demand for labor by a firm approximately unitary elastic. Finally, we calibrated the numerical model via the safety production function ($s = s_0 + s_1\pi + s_2\pi^2$) using the initial conditions of (1) a 1 in 1,000 average probability of serious injury, (2) a 0.40 proportion of firms without incremental workplace hazards, and (3) a 1 in 1,000 maximum probability of serious injury. Appendix 3A provides a more complete explanation of calibration and the selection of initial conditions and production function parameters.

We also followed the approach in Chapter 3 of adding a payment representing fixed costs so the marginal firm earns only a normal rate of return in long-run equilibrium. The value we selected for fixed cost, c, causes the marginal employer to have a value for μ, denoted $M(\pi^m)$ equal to 0.1. Thus, the initial values observed for μ range from 0.1 to 1.

To complete our numerical description of a firm's shareholder wealth we had to specify the relationship between injury rates and expected fines for OSHA violations. In the initial simulations and the calibration process we assumed the situation prior to OSHA and, therefore, used zero expected fines at all levels of safety. We discussed the post-OSHA situation in the chapter body.

Input and Output Markets

We assumed that both input and output markets are competitive. Firms purchase capital and safety equipment at fixed prices of $1. We implicitly assumed both inputs are produced in constant cost industries, implying the prices of safety and capital equipment are invariant to the particular changes in the economic environment we considered in the numerical simulations.

We describe the output market by a unitary elastic demand function. We initially set the price of output at $1. Later, when we investigate OSHA–WC interactions, we allowed the price of output to adjust. Hence, changes in labor costs have direct effects and indirect effects on hedonic labor market equilibrium through their impact on product price. Likewise, changes in workplace safety alter WC insurance prices which then directly and indirectly affect hedonic labor market equilibrium. We discuss the insurance market in the chapter body.

Finally, our numerical simulation used a fixed number of workers and firms. Because labor is measured in terms of dollars of net return to the firm payroll expenditures also equal total labor input. To reproduce labor market equilibrium prior to OSHA we varied the number of workers in the model until total payroll equaled the actual payroll in the manufacturing sector of the United States in 1972. The relevant number of establishments in our model includes both producing firms and potential entrants. To create the pre-OSHA situation we increased the number of firms observed in the U.S. manufacturing sector during 1972 to reflect an assumption of a 10 percent pool of potential entrants. Thus, the number of firms producing in our initial simulation, which is labor market equilibrium prior to the introduction of OSHA and any revisions in WC, equals the number of manufacturing firms actually in existence in 1972.

NOTES

[1]Viscusi (1979, 1983) provides background.

[2]OSHA has used both performance and specification standards. Performance standards set maximum levels of exposure to particular hazards, whereas

specification standards mandate specific types of safety programs. One can view performance and specification standards as regulating different aspects of the production function for safety. Although the model we simulated is based on performance standards, the examination of specification standards would be identical. With specification standards expected fines are a decreasing function of the quantity of safety equipment.

[3]Assuming the distribution of violations among nonserious, serious, and willful categories is constant at their initial levels and the average number of citations remains 6.8, an average fine of $1,107 (1993 dollars) implies a doubling of all fines from the levels set in 1973. To represent even greater enforcement, we varied v_1 to produce an average fine of $3,980. Again we assumed the distribution and the number of citations remained at 1973 levels, but that OSHA raised fines for nonserious violations 10 times higher than initially and imposed fines for serious and willful violations of $3,255 and $32,545, which were the maximum legislated limits (1973 nominal limits were $1,000 and $10,000, respectively). Finally, to represent extremely stringent enforcement efforts, we varied v_1 to produce an average fine of $22,329, which duplicated the previous situation of moderate enforcement, except fines for nonserious violations averaged $3,255, the legal maximum (the 1973 nominal maximum was $1,000).

[4]Based on OSHA data (U.S. Department of Labor 1973), the probability of death for a worker in the manufacturing sector in 1972 was 0.00007. However, the Society of Actuaries reported, for at least in a segment of the manufacturing industry we considered, that the mean *incremental* probability of death was 0.001, which suggests that in the eyes of OSHA authorities and the WC system only about seven percent of all worker deaths are truly job-related. As Viscusi (1986c) noted the WC system does not cover work-related health impairments as effectively as injury risk because of the difficulty in monitoring causality. Because the Society of Actuaries investigated only very hazardous occupations we used the higher 10 percent rate to represent the probability of receiving WC benefits if seriously injured.

[5]As noted earlier, initially OSHA fined firms only $553 on average for violating safety and health standards. Normal randomness would most likely mask the 0.3 percent reduction in the frequency of injury we calculated. Further, 0.3 percent probably represents an upper bound estimate of OSHA's initial impact. Although we created an average fine of $553, we assumed OSHA inspected 10 percent of all establishments. In 1972 OSHA inspected only 36,100 out of a total 4,946,300 establishments covered by the Act, less than a 1 percent rate. In the second half of 1972 OSHA focused 90 percent of all inspection activities in three major industry categories — construction,

manufacturing, and maritime. Even with inspection targeting OSHA visited only 0.4 percent of all construction establishments and only 2.7 percent of all manufacturing establishments during the period. Therefore, the assumption of a 10 percent inspection rate overstates the actual probability of inspection. When examining the financial incentives to comply with OSHA standards Viscusi (1986b, 1986c), for instance, uses only a 0.5 percent chance of inspection and determines the resulting expected fine to be only 83 cents per worker (1993 dollars).

Additionally, from our numerical simulation results we calculated total OSHA fines in the manufacturing sector to be $17,743,559 (1993 dollars) which is 0.003 percent of payroll. In actuality OSHA imposed total penalties of only $10,157,300. In the second half of 1972, when OSHA compiled information on type of establishments, manufacturing firms received 14 percent of all citations issued. Assuming the average fine in the manufacturing sector was the same as the average fine for all types of establishments, OSHA penalized manufacturing firms $1,422,023, which is 0.00025 percent of payroll. Assuming the manufacturing sector accounted for all violations, OSHA penalized manufacturing firms 0.002 percent of manufacturing payroll. Thus, we use an initial level of OSHA enforcement that is between 1.5 and 12 times greater than what actually existed.

Finally, the results in Table 4-2 may also overstate OSHA's initial impact because we assumed safety and health regulators correctly identified and cited employee hazards. Critics claim OSHA standards do not address health and safety issues and, even if completely implemented, would not significantly reduce industrial accidents and diseases. Viscusi (1986b, 1986c) discusses the problems of OSHA specification standards and the nontechnological aspects of many accidents (for example, accidents caused primarily by worker actions). If OSHA's critics are correct, and at least some of OSHA's activities are unrelated to safety, then OSHA's impact is lower than what we computed.

[6]As we will see more clearly in the next section, WC benefits *increase* the average frequency of injury. As a result, Table 4-2 shows that OSHA reduces injuries slightly more when the average probability of injury is small in the first place. Practically speaking, WC reduces the effectiveness of OSHA at the time when, some might argue, workers most require safety and health standards. However, the reduction in OSHA's effectiveness due to WC is small.

[7]Remember with the constant partial elasticity of substitution production function we used safety equipment augments labor productivity. As long as OSHA enforcement activities increase the total usage of safety equipment average wages increase.

[8]The average frequency of injury is

$$\bar{\pi} = \int_0^{\pi^m} \pi f(\pi, v, i)\, d\pi,$$

where $\bar{\pi}$ ≡ the average frequency of injury, and
$f(\pi)$ ≡ the equilibrium distribution of workers with respect to π.

Totally differentiating the expression with respect to v generates

$$\frac{d\bar{\pi}}{dv} = \int_0^{\pi^m} \pi \frac{\partial f(\pi, v, i)}{\partial v} d\pi + \pi^m f(\pi^m, v, i) \frac{d\pi^m}{dv}.$$

We can interpret the first term on the right-hand side as the impact of OSHA enforcement activities on the equilibrium distribution of workers and the second term as the impact of OSHA on the maximum level of risk. To separate the two effects in our numerical simulation results we assumed the distribution of workers within a 0.00025 risk interval is approximately uniform. At the original equilibrium $f(\pi^m, v, i) =$ 1172900/0.00025 = (4.6916×10^9), where 1172900 is the number of workers initially in the most dangerous jobs ($\pi >$ 0.00275). Using 1172900 for the density of workers at π^m and the numerical results for $d\pi^m/dv$ we can directly calculate the impact on injuries caused by π^m changing. We calculated the impact of OSHA on the equilibrium distribution of workers, what we have called the sorting effect, by subtracting the impact due to changing π^m from the total change in injuries.

[9]The implied elasticity of π with respect to i seems to be much smaller than suggested by the econometric literature; one possible reason for the difference is miscalculation of injuries as work related in the data used in the econometric research cited.

[10]Remember there is an upward sloping long-run output supply curve caused by firm heterogeneity and a downward sloping long-run product demand curve.

[11]We altered the two programs prior to examining changes caused by increasing OSHA fines or WC benefits. In each case we recalibrated the model to produce the initial conditions listed in Table 4-1.

5 ASYMMETRIC INFORMATION PROBLEMS IN WORKERS' COMPENSATION INSURANCE

We have emphasized that workers' compensation insurance (WC) is the most important public policy influencing workplace safety in the United States. We now examine the economic and policy implications of asymmetric information on the part of workers and firms concerning injury severity and compensability under WC. Concern over the difficulty in determining the information content of workers' compensation claims is underscored in a recent *Economic Report of the President* (1987, p. 197), which noted that

> a growing body of research has found that workers' compensation benefits have unfavorable effects on safety. Higher benefits appear to increase both the frequency of work injuries and the number of compensation claims filed. One explanation for the positive connection is the claim effect. Even if actual injuries remain constant, workers are more likely to file claims when benefits are higher, thereby producing more reported injuries.

Here we use our numerical simulation model to clarify the quantitative relationships among the parameters of WC, the number of insurance claims, and worker safety. We demonstrate that three frequently discussed reforms — more generous benefits to injured workers, more accurate categorization of insurance claims, and more complete experience rating of WC premiums — will each slightly reduce total work-related injuries but substantially raise the total number of workers' compensation insurance claims.

LABOR MARKET EQUILIBRIUM

The recurring theme in the economics of workplace safety is heterogeneity — employers differ in their costs of making the workplace safer and workers differ in their willingness to accept the risk of work-related injuries. The result of firm–worker heterogeneity is that no single level of job safety is best and that in the process of matching jobs and workers a multiplicity of safety levels are created. Labor market equilibrium, described by the hedonic wage function, is the long-run relationship between wages and the likelihood of a work-related injury that equates the quantities of labor demanded and supplied at the various levels of injury risk (Rosen 1974, 1986).

Workers

The primary difference between our simulation model acknowledging asymmetric information in workers' compensation and the approach in Chapter 4 comparing OSHA and workers' compensation safety effects is that here we consider that injuries are difficult to classify as work related or not. In particular, we now let the worker have the potential for any of three work-related injures: a nonimpairing injury (NII) that is a minor injury having no effect on workplace productivity, a permanent partial disability (PPD), that is a moderate injury somewhat reducing workplace productivity and earning capacity, and a permanent total disability (PTD).[1] In our simulation model all injuries happen in the workplace but there is imperfect screening that lets some workers with minor, nonimpairing, injuries receive WC benefits if they apply and denies benefits to some moderately or severely injured workers because employers have questioned the severity or work relatedness.

Applying for WC Benefits: Background. When making their labor supply decisions workers take into account WC benefits, which are not taxed, the probability of receiving WC benefits once they apply for benefits, and whether it is worthwhile to apply for WC benefits. Because some workers exaggerate NII or PPD injuries into permanent total disabilities when applying for WC, insurers try to determine the extent of an injury to avoid overcompensating a worker's income loss.

The process that classifies injuries as NII, PPD, or PTD is imperfect because workers have the best information on the true cause and extent of their injuries, workers have the financial incentive to exaggerate injury severity, and firms have the financial incentive to classify injuries as unrelated to work or less severe than the injuries are. One study of Social Security Disability Insurance (SSDI) found that a second evaluation team would reclassify one-fifth of eligibility rulings (Parsons 1984). Providers of WC face a more complex decision process than in the case of SSDI because injuries compensable under WC need to have occurred in the course of employment.

A worker filing an exaggerated WC claim must consider that permanent total disability benefits are usually less than wages and that the filing process has costs. For instance, a rejected applicant can have lower wages when reentering the labor force because human capital and productivity may decline when out of the labor force during the application process, employers view a period of nonwork as an indication of permanent impairment of potential future work, and employers may interpret time out of the labor force as signaling weak labor force attachment. Evidence is that workers denied disability payments under SSDI suffered a 10 percent reduction in wages when they later reentered the labor force compared to similarly injured workers who never applied for insurance benefits (Halpern and Hausman 1986).

In addition to the indirect costs of reduced future earning potential a worker considering filing an exaggerated claim for WC will take account of possible guilt feelings, so-called welfare stigma (Moffitt 1983), and how filing a total disability claim affects the possibility of receiving a partial disability award. Our numerical simulation model also incorporates the fact that workers denied WC payments can receive other social insurance benefits related to low income.

Expected Utility. A worker's expected utility when there are three types of work-related injuries that may be incorrectly represented by workers or incorrectly classified by insurance providers is

$$
\begin{aligned}
\bar{u} &= (1-\pi_1-\pi_2-\pi_3)w^{1-\alpha}l^{\tau} \\
&+\pi_1\left[\max\left\{\begin{array}{l} w^{1-\alpha}l^{\tau}, \\ \psi_1\beta_1(wi)^{1-\alpha}\tilde{l}^{\tau}+(1-\psi_1)\left(\psi_1(w(\tilde{d}+(1-d)i))^{1-\alpha}l^{\tau}+(1-\psi_1)(w\tilde{d})^{1-\alpha}l^{\tau}\right)\end{array}\right\}\right] \\
&+\pi_2\left[\max\left\{\begin{array}{l} \psi_2(w(d(1-d)i))^{1-\alpha}l^{\tau}+(1-\psi_2)(w\tilde{d})^{1-\alpha}l^{\tau}, \\ \psi_1\beta_2(wi)^{1-\alpha}l^{\tau}+(1-\psi_1)\left(\psi_2(w(\tilde{d}+(1-d)i))^{1-\alpha}l^{\tau}+(1-\psi_2)(w\tilde{d})^{1-\alpha}l^{\tau}\right)\end{array}\right\}\right] \\
&+\pi_3\left[\psi_2\beta_3(wi)^{1-\alpha}\tilde{l}^{\tau}+(1-\psi_2)\left(\psi_2\beta_3(w(1-d)i)^{1-\alpha}\tilde{l}^{\tau}+(1-\psi_2)\beta_3 y_a^{1-\alpha}\tilde{l}^{\tau}\right)\right]. \qquad \textbf{(5.1)}
\end{aligned}
$$

$\bar{u}$ is expected utility; π_1,π_2, and π_3 are the probabilities of a nonimpairing injury, a permanent partial disability, and a permanent total disability; w is the market wage, which is a function of the average probability of a work-related injury (π); l is leisure if in the labor force; ψ_1 and ψ_2 are the probability of receiving WC benefits for an exaggerated claim and the probability of receiving WC benefits for a genuine claim; β_1 and β_2 are the fractions of pre-injury utility, income held constant, if falsely receiving PTD benefits; i is the fraction of income replaced via WC; $\tilde{l}$ is leisure if out of the labor force; $\tilde{d}$ is the fraction of wages a worker earns if denied WC benefits; d is the fraction of wages a PPD worker earns if receiving WC benefits; β_3 is the fraction of pre-injury utility, income held constant, if a PTD occurs; y_a is the income received from public and private assistance if denied WC benefits; α is the degree of risk aversion parameter; and τ is a parameter determining the utility of leisure.

We make the background assumptions that workers' compensation benefits replace a fraction of foregone wages, that PPD workers stay in the labor force, and that NII workers denied WC benefits earn the same as PPD workers denied WC benefits because employers are unable to distinguish among workers denied WC benefits. Our final background assumptions differing from previous chapters are that when maximizing utility in (5.1) workers choose jobs based on the average level of injury risk, $\bar{\pi} = \pi_3(f_1+f_2+1)$, where $f_1 \equiv \pi_1/\pi_3$ and $f_2 \equiv \pi_2/\pi_3$ are each assumed to be constant across employers. Because the relative probabilities of minor and moderate injuries do not vary across workplaces, workers can be viewed as choosing where to work based on the probability of a PTD (π_3). Interpersonal differences in risk aversion, which are metered by the parameter α in the expected utility function in (5.1), influence the perceived marginal gain from higher wages on more dangerous jobs and the worker's welfare loss from a work-related injury, which causes the optimal amount of workplace safety to differ across people.[2] The expected utility function we used (5.1) then captures the dual decisions concerning whether to work for pay and whether to file an exaggerated claim for PTD benefits if suffering a nonimpairing or moderate injury.

Employers

Firms choose the combination of workers, capital, and expenditures on workplace safety that maximize their expected profits. When deciding how much to spend on safety enhancing measures employers consider the possible economic benefits of a safer workplace, such as greater output,[3] lower pay for their workers, and lower workers' compensation insurance premiums. To keep our simulations with a multiplicity of injury classes simple we again followed the assumption of our earlier chapters in having firms choose the level of a generic device, safety equipment, that represents all safety enhancing expenditure firms make. A firm's expected profit then depends on the price of its product and its choices of safety equipment, capital, and its (effective) number of workers, n_e, which is less than its total labor force, n.[4]

Expected Profit. In general algebraic notation including input and fixed costs a firm's expected profit is

$$\bar{g} = p_o Q(S(\pi), k, n_e; \mu) - wn_e - P_b(\pi, a_i, a_2, \psi_1, \psi_2, i)wn \\ -p_s S(\pi) - p_k k - c. \tag{5.2}$$

$\bar{g}$ is expected profit; $Q(\cdot)$ is output, which depends positively on safety expenditures (s), capital (k), and effective labor (n_e); the safety production function, $s = S(\pi)$ with $S' < 0$, acknowledges that safety expenditures and workplace injury risk are inversely related; and μ is a parameter for heterogeneous efficiency in the joint production of safety and output for sale ($\partial^2 Q/\partial s \partial \mu > 0$). The price of WC insurance as a fraction of labor costs, P_b, depends positively on the fraction of a firm's NII injured workers expected to file for WC benefits (a_1), the fraction of a firm's PPD workers expected to file exaggerated claims for PTD benefits (a_2), the probability a worker receives WC benefits for an exaggerated claim (ψ_1), the probability a worker receives WC benefits for a genuine claim (ψ_2), and i, the income replacement rate under WC. Finally, p_o is the price of output (q), p_s is the price of a unit of safety equipment, p_k is the price of a unit of capital, and c is total fixed cost of production. In completing the firm's objective function, (5.2), we used the same output and safety production functions as in earlier chapters.

Output Production Function. We again used a constant partial elasticity of substitution production function in our numerical simulations that is

$$q = a\left[\delta(\mu s)^{-\rho} + \gamma k^{-\rho} + (1 - \delta - \gamma)n_e^{-\rho}\right]^{-\upsilon/\rho}, \tag{5.3}$$

where a is the efficiency parameter, δ and γ are distribution parameters, ρ determines the elasticity of substitution in production $\sigma = 1/(1+\rho)$, and υ is a parameter determining returns to scale. As noted earlier we assumed a technology where safety is a jointly productive input in that safety enhancements (s) increase output by reducing injuries to workers.

The productivity of safety expenditures in output production is captured by the parameter μ in the production function (5.3). In simulating the effects of workers' compensation insurance under asymmetric information and imperfect state verification of injuries employer heterogeneity appears as a uniform random distribution of μ across firms in the interval 0 to 1. Firms with the largest values of μ are the most efficient providers of safety to their workers and in turn have the safest workplaces.

Safety Production Function. The other structural equation in firms' expected profit equation (5.2) is the safety production function describing how safety expenditures influence the degree of injury risk of workers, $S(\pi)$. Little is known empirically about the link between safety expenditures and workers' job-related injury risk. We used a noncontroversial specification for our simulations that has safety expenditures quadratic in worker injuries,

$$s = s_0 + s_1 + s_2\pi^2. \tag{5.4}$$

As in Chapters 3 and 4 we calibrated the simulation model with the safety production function using a base case labor market outcome to infer values of s_0, s_1, and s_2 in (5.4). The firm then selects its expenditures on labor, capital, and safety so as to maximize expected profits (5.2), which depend on the output production function in (5.3) and the safety production function in (5.4).

Labor Market Equilibrium

When workers and firms are heterogeneous concerning a willingness to accept workplace injury risk (α) and the costs of creating a safer working environment (μ) there is a matching process between the quantities of labor supplied and demanded across the joint wage and injury risk spectrum. The hedonic equilibrium labor market locus shows combinations of wages (w) and average injury probabilities (π) that clear the labor market under joint worker–firm heterogeneity.

As in the models developed in Chapters 3 and 4 the hedonic labor market equilibrium satisfies the equation

$$L(\alpha)\left|\frac{d\alpha}{d\pi}\right|d\pi = N(\mu)F(\mu)\left|\frac{d\mu}{d\pi}\right|d\pi. \tag{5.5}$$

$L(\cdot)$ is the density function of workers with respect to their risk aversion parameter (α), where α takes on values between some maximum and minimum with equal probabilities; $N(\cdot)$ is the demand for labor by an individual firm, which solves the firms three necessary conditions for maximum profit; and $F(\mu)$ is the density function of firms with respect to their efficiency of safety production, where $0<\mu\leq 1$ with equal probabilities. Appearing on the left-hand side in the hedonic labor market equilibrium equation, (5.5), is the number of workers supplying labor to employers at each risk (π) level, and appearing on the right-hand side of (5.5) is the quantity of labor demanded.[5]

The differentials in the equation for hedonic labor market equilibrium, $d\alpha/d\pi$ and $d\mu/d\pi$, describe how firms and workers match at the various degrees of exposure to work-related health hazards. There are three first-order conditions for the firm's maximum profit. We totally differentiated the three necessary conditions for maximum profit in (5.2)–(5.4) with respect to μ and π to solve for the equilibrium pairings of firms and workplace safety levels, $d\mu/d\pi$. Next, we re-arranged equation (5.5), the necessary condition for labor market equilibrium, to solve for $d\alpha/d\pi$, how workers sort against levels of injury risk. The differential equations of interest have the general expressions

$$\frac{d\mu}{d\pi} = M'(\pi,\ \alpha,\ \mu,\ w, i, a_1, a_2, \psi_1, \psi_2, P_b,\ W'), \tag{5.6}$$

and

$$\frac{d\alpha}{d\pi} = A'(\pi,\ \alpha,\ \mu,\ w, i, a_1, a_2, \psi_1, \psi_2,\ P_b), \tag{5.7}$$

where P_b represents the parameters of the function describing the pricing of workers' compensation insurance. Finally, we took the single necessary condition for individual expected utility maximization from (5.1) and totally differentiated with respect to π, which let us solve for how injury risk (positively) influences wages in labor market equilibrium

$$\frac{dw}{d\pi} = W'(\pi,\ \alpha,\ w, i, \psi_1, \psi_2).^6 \qquad \textbf{(5.8)}$$

NUMERICAL SIMULATION DETAILS

Unlike early research using hedonic equilibrium, which developed closed form solutions for market outcomes by abstracting from uncertainty or endogenous decisions by firms (Tinbergen 1956, Sattinger 1977), we have a full supply and demand framework under uncertainty. Because we want to examine how wages and industrial-injury risk are influenced by the economic environment, including policies such as workers' compensation insurance, we had to consider the endogenous demand for labor by firms and the decision to supply labor by workers in an environment where danger of damage to health status is stochastic. Even though they have been stripped of unnecessary details the three differential equations describing labor market equilibrium are still too complex for an analytic solution and examination of equilibrium disturbances via comparative statics. As in Chapters 3 and 4 we solved numerically for hedonic labor market equilibrium in (5.6)–(5.8) using realistic numerical values for the system's parameters and boundary conditions.

Starting Parameter Values

We built our numerical simulations by first parameterizing the representative worker's expected utility in (5.1), the representative firm's production function (5.3), and the probability density functions for workers' risk aversion heterogeneity and firms' safety efficiency heterogeneity, which are $L(\alpha)$ and $F(\mu)$ in the hedonic equilibrium equation, (5.5). We chose numerical starting values that make the model replicate equilibrium in the labor market for manufacturing workers in the early 1970s, which was examined extensively empirically by Thaler and Rosen (1975). Our initial hedonic equilibrium is for the U.S. manufacturing industry just before the National Commission on State Workmen's Compensation Laws recommended numerous reforms ultimately adopted by state WC programs (Oi 1973, Darling-Hammond and Kniesner 1980). From our baseline simulation we then examined how changes in the WC system affect the frequency of workplace injuries, the number of WC claims filed, and the number of WC claims paid. Because our starting parameter values are generally the same as in simulations in earlier chapters we describe now only the parameters of WC relating to the asymmetric information and imperfect state verification of work-related injuries.

The Workers' Compensation Insurance System

There are six dimensions of Workers' Compensation to consider concerning imperfect verification of the severity and compensability of employee injuries: (1) the proportion of foregone income that WC replaces, (2) the probability that a worker filing an exaggerated claim gets WC benefits, (3) the probability that a worker filing a genuine claim receives WC benefits, (4) the fraction of pre-injury earnings that a worker receives when returning to work after suffering a permanent partial disability injury and having been denied WC benefits, (5) the fraction of pre-injury earnings that a worker receives after a permanent partial disability injury if the worker does not file for workers' compensation benefits, and (6) the income a worker receives from other public assistance programs if denied WC benefits.

No single number describes the income replacement rate in U.S. workers' compensation due to the diversity of programs across states and because benefit levels differ according to injury severity and worker income (Viscusi and Moore 1987). The mode level of income replacement is about two-thirds of gross average weekly earnings for permanent total disability injuries (Darling-Hammond and Kniesner 1980). However, most states tie maximum weekly benefits to the state's average wage rate so that injured workers often receive less than two-thirds their pre-injury earning capacity under WC. In light of the interstate dispersion of replacement rates we chose the median after tax income replacement rate for married men, which was 50 percent in the late 1970s (Johnson and Wai 1980).

The likelihood of receiving workers' compensation benefits if a worker files an exaggerated claim, ψ_1, and the likelihood of receiving WC benefits if a worker files a genuine claim, ψ_2, represent the degree of imperfect state verification of work-related injures in our numerical simulation model. Under completely accurate screening $\psi_1=0$ and $\psi_2=1$, so that every exaggerated claim would be denied and every genuine claim would be paid. In reality, some persons injured at work are denied WC benefits that would be paid if the true cause and severity of their injuries were known, and some workers receive WC benefits that would be denied if the true source and severity of their injuries were known. The degree of imperfect state verification of disability has been extensively studied in the case of Supplemental Social Security Disability Insurance (SSDI) but not workers' compensation insurance. As a starting point we applied the parameters of claims screening accuracy of SSDI, which are $\psi_1=0.34$ and $\psi_2=0.7$, or 34 percent of exaggerated claims are paid and about 70 percent of genuine claims are paid (Parsons 1984).

We selected values for the parameters describing the fraction of previous wages a worker earns if denied WC benefits, $\tilde{d}$, and the fraction of wages that a worker earns if he or she suffers a PPD and does not apply for WC benefits, d, from research on SSDI. Econometrically estimated wage functions including health status and the application for WC benefits as regressors suggest the typical worker with a health impairment has wages that are about 36 percent less than an otherwise similar healthy worker, and an unsuccessful applicant for SSDI benefits has wages that are 10 percent lower than a nonapplicant, ceteris paribus (Halpern and Hausman 1986). Because comparable values do not exist in the literature for WC we borrowed from the research on SSDI, which gives us starting values of d=0.65 and $\tilde{d}$ =0.60, so that initially we assumed a worker who is denied WC benefits can earn 60 percent of his or her previous wage and a PPD worker who does not apply for WC can earn 65 percent of the previous wage.

The last parameter we mention is the income that a worker receives if he or she suffers a permanent total disability but is denied workers' compensation insurance benefits due to imperfect state verification that misclassifies the injury as unrelated to work. During the early 1970's maximum annual welfare (AFDC) payments ranged from about $700 in Mississippi to more than $4000 in Massachusetts, New York, and New Jersey (Reynolds 1974). In light of the range of potential AFDC benefits and the wages paid on jobs with minimal injury risk we chose $1000 (as the public assistance payment available to workers suffering a permanent total disability but, due to imperfect verification of the source and severity of their injuries, are denied workers' compensation benefits.[7] In practicality, public assistance of $1000 ($3,724, 1993 dollars) for an uncovered but injured worker is about a 90 percent decline in income from the pre-injury level and about one-fifth of the average recipient's WC benefits, so that in our simulations there is a substantial economic loss for workers who are injured but are not compensated under WC.

Summary of Starting Parameters

Because the parameters other than the values describing the WC system under asymmetric information and imperfect state verification of injuries used in our simulation model have been discussed in Chapters 3 and 4 we will not repeat their quantitative justification here. For the reader's convenience all numerical parameters and initial conditions that are the basis of our initial hedonic equilibrium locus appear in Table 5–1.

Table 5–1. Starting Parameter Values

I. Workers

$\alpha \equiv$ degree of risk aversion ($0.8 \leq \alpha \leq 1.0$)
$a_1 \equiv$ fraction of NII workers filing for WC (varies by α)
$a_2 \equiv$ fraction of PPD workers filing as PTD (varies by α)
$l \equiv$ leisure time of a worker (=6760)
$\tilde{l} \equiv$ leisure time of a person not in the labor force (=8760)
$\tau \equiv$ parameter determining the relative utility of leisure (=0.8)
$\beta_1 \equiv$ relative utility from a successful false PTD claim if NII (=0.94)
$\beta_2 \equiv$ relative utility from a successful false PTD claim if PPD (=0.85)
$\beta_3 \equiv$ relative utility if PTD (=0.10)
efficiency units of labor per worker (=10,000)

II. Firms

$\mu \equiv$ relative efficiency of safety enhancements ($0<\mu\leq 1$)
$\delta \equiv$ distribution parameter on safety in the production function (=0.1)
$\gamma \equiv$ distribution parameter on capital in the production function (=0.2)
$\upsilon \equiv$ returns-to-scale parameter in the production function (υ=0.1)
$\rho \equiv$ elasticity of substitution parameter (=0.01)
$c \equiv$ fixed cost of production (=$6,845,000)
$s_0 \equiv$ safety production function parameter (=169,000)
$s_1 \equiv$ safety production function parameter (=-4,705,000)
$s_2 \equiv$ safety production function parameter (=-10,219,000)

III. Product and Input Markets

$p_k \equiv$ price of a unit of capital (=$1)
$p_o \equiv$ price of output (=$1)
$p_s \equiv$ price of a unit of safety enhancements (=$1)
$\eta \equiv$ product price elasticity (=-1)
$n \equiv$ number of workers (= 17,259,000)
number of firms (= 356,344)

IV. Workers' Compensation Insurance

$p_b \equiv$ price of WC insurance per unit of coverage (varies by risk level)
$i \equiv$ WC income replacement rate (=0.50)
$\psi_1 \equiv$ the probability of WC benefits for an exaggerated claim (=0.34)
$\psi_2 \equiv$ the probability of WC benefits for a genuine claim (=0.7)
$y_a \equiv$ public assistance if denied WC benefits (=$1000; $3,724, 1993 dollars)

Table 5-1. (Continued)

$d \equiv$ fraction of pre-injury wages earned if PPD and no WC denial (=0.65)
$\tilde{d} \equiv$ fraction of pre-injury wages earned if denied WC benefits (=0.60)
degree of experience rating = 33 percent

V. Worker–Firm Heterogeneity
$L(\alpha) = 8.6295\times10^8$ if $\alpha_{min} = 0.8 \leq \alpha \leq 1.0 = \alpha_{max}$; $L(\alpha)=0$ otherwise
$F(\mu) = 356{,}344$ if $0<\mu\leq 1$; $F(\mu) = 0$ otherwise

VI. Initial Conditions
$\bar{\pi} \equiv$ average frequency of injury (=0.00975)
$W(\pi^{min}) \equiv$ wage rate in the minimum risk employer (=$10,000; $37,242, 1993 dollars)
$\pi^{min} \equiv$ minimum level of injury risk (=0.006)
$\pi^{m} \equiv$ maximum level of injury risk (=0.015)
$f_1 \equiv$ NII injury risk/PTD injury risk (=1.0)
$f_2 \equiv$ PPD injury risk/PTD injury risk (=1.0)
$M(\pi^{m}) \equiv \mu$ of the employer with the maximum injury risk (=0.1)
$M(\pi^{min}) \equiv \mu$ of the employer with the minimum injury risk (=0.7)

NUMERICAL SIMULATION RESULTS

Figure 5.1 depicts the simulated hedonic labor market equilibrium prior to our calculations of how workers' compensation reforms influence outcomes of interest, including wage levels and average injury risk.[8]

Because some workers exaggerate the true extent of their injuries and because WC providers deny some valid claims, total PPD plus PTD claims paid are lower than the actual number of PPD and PTD injuries. In our baseline numerical simulation the actual frequency of a debilitating injury (a PPD or PTD) is 6.5 per 1000 workers but the number of claims filed is 15 percent greater (7.5/1000) because imperfect monitoring of insurance claims causes some NII workers to apply for WC benefits. In addition, the application rate for PTD benefits (5.24/1000) is 60 percent greater than the actual number of PTD injuries (3.25/1000) because imperfect state verification also causes

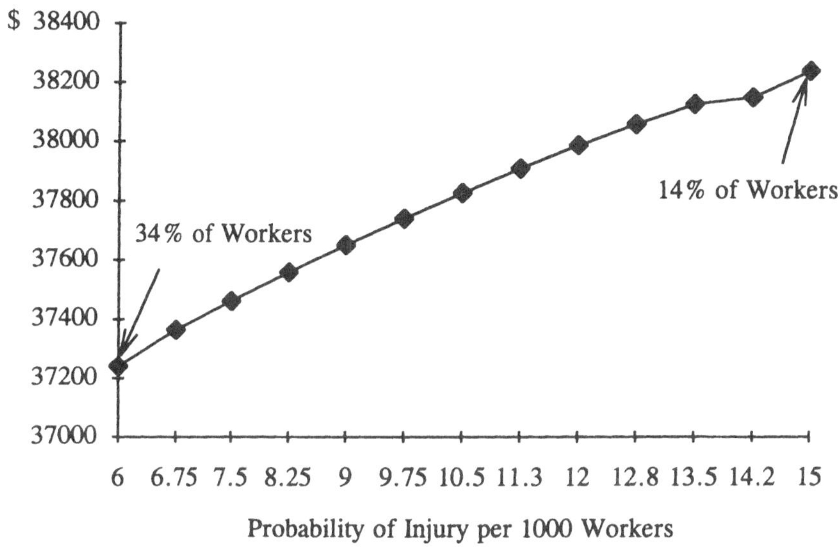

Figure 5.1 Baseline Hedonic Equilibrium Wage Function (1993 dollars)

some workers with PPD's to file exaggerated claims for PTD's. The result of asymmetric information and imperfect state verification of the source and severity of worker injuries is that the number of PPD claims filed under WC (2.25/1000) understates the number of PPD injuries (3.25/1000), and the number of PTD claims filed (5.24/1000) overstates the number of PTD injuries (3.25/1000). Our baseline simulation results underscore that injury rate estimates based on WC claims filed understate the degree of workplace safety, and injury rate estimates based on the number of WC claims paid overstate the degree of industrial safety.

The Effects of Workers' Compensation Benefit Levels

To examine how workers' compensation benefits influence labor market outcomes we simulated the effects of increasing the earnings replacement rate by 10 percent, 20 percent (the reform suggested by the National Commission on State Workmen's Compensation Laws), and 100 percent (which is complete income replacement). We also simulated how eliminating workers' compensation insurance would affect wages and workplace safety. Moving to full income replacement or no income replacement are extreme changes that require a complete structural model of WC and the labor market such as our numerical simulation model. We would be hesitant to accept forecasts of the effects of extreme changes in the WC system from the typical skeletal econometric model of hedonic labor market equilibrium. Here is another example of how numerical simulation can complement econometric research to sharpen our knowledge of how public policy and other changes in the economic environment influence labor market outcomes, particularly how WC affects actual versus reported industrial injuries.

The results in Table 5-2 illustrate a negative effect of benefit generosity on injury rates. Higher income replacement rates give firms financial incentives to improve workplace safety via the pricing mechanism of workers' compensation insurance, which translates greater generosity for injured workers into higher WC premiums for employers. The increase in workplace safety due to greater WC benefit generosity is small, though. A 100 percent increase in the income replacement rate raises the injury rate by about two percent. In our four benefit generosity simulations the injury rate/benefit generosity elasticity is -0.01 to -0.03.

Although more generous WC income replacement is safety enhancing, applications for WC and claims paid rise with WC generosity. In Table 5-2 there is a simulated response of four percent more applications to a 20 percent increase in benefit generosity. The elasticity of PTD applications with respect to WC replacement is about 1.0 and the elasticity of PPD claims filed to WC income replacement is about -1.0 to -2.4. Greater WC generosity increases the incentive for NII workers to exaggerate claims, which causes the overall increase in claims. However, the distribution of claims is also altered by greater WC generosity as is there more exaggeration of PPD into PTD claims.

Table 5-2. The Effects of WC Benefits and Experience Rating on Probability of Injury, Claims Filed, and Claims Paid [Percentage Changes (Elasticities) from the Baseline Simulation in Figure 5.1 and Table 5-1]

Benefits

Income Replacement Rate (% change)	Injuries	Application Rate — WC Benefits	Application Rate — PTD Benefits	Application Rate — PPD Benefits	Frequency of WC Claims Paid	Frequency of PTD Claims Paid	Frequency of PPD Claims Paid
0 (-100%)	1.10% (-0.01)	-100% (1.0)	-100% (1.0)	-100% (1.0)	-100% (1.0)	-100% (1.0)	-100% (1.0)
0.55 (10%)	-0.28 (-0.03)	2.36 (0.23)	9.53 (0.95)	-14.31 (-1.43)	2.16 (0.21)	5.64 (0.56)	-1.33 (-0.13)
0.60 (20%)	-0.56 (-0.03)	3.97 (0.20)	26.64 (1.33)	-48.74 (-2.44)	4.57 (0.22)	15.86 (0.79)	-6.75 (-0.34)
1.00 (100%)	-2.04 (-0.02)	27.43 (0.27)	82.23 (0.82)	-100 (-1.0)	22.93 (0.23)	48.83 (0.49)	-3.06 (-0.03)

Table 5-2. *(Continued)*

Experience Rating

Degree of Experience Rating (% change)	Injuries	Application Rate — WC Benefits	Application Rate — PTD Benefits	Application Rate — PPD Benefits	Frequency of WC Claims Paid	Frequency of PTD Claims Paid	Frequency of PPD Claims Paid
0.363	-0.15	-0.11	-0.04	-0.28	-0.12	-0.08	-0.15
(10%)	(-0.01)	(-0.01)	(*)	(-0.03)	(-0.01)	(-0.01)	(-0.02)
0.396	-0.30	-0.22	-0.08	-0.56	-0.23	-0.17	-0.30
(20%)	(-0.02)	(-0.01)	(*)	(-0.03)	(-0.01)	(-0.01)	(-0.02)
1.0	-0.75	-0.56	-0.20	-1.39	-0.59	-0.42	-0.76
(200%)	(*)	(*)	(*)	(-0.01)	(*)	(*)	(*)

Note: Table 5–1 lists the starting parameter values used in the above simulations. All frequencies, probabilities, and rates are relative to the total labor force.

*Less than 0.005

The inelastically positive simulated response of reported work-related injuries to benefit generosity in Table 5-2 has also appeared in econometric research, which has found that states with higher WC benefits have higher injury rates (Butler 1983; Butler and Worrall 1983; Chelius 1974, 1982; Ruser 1985). Imperfect verification of the cause and severity of injuries makes true injuries and injuries reported to WC diverge. Econometric research often relies on reported workplace injuries. However, WC changes not only the costs and benefits of safety enhancements to firms but also changes the relative payoffs to workers from reporting and exaggerating injuries to the WC authorities. In our simulation model WC benefit increases make reported injuries rise substantially while actual workplace injuries fall slightly. Our numerical simulation model indicates that the so-called reporting effect, where reported injuries diverge from actual injuries, can lead to incorrect inferences of the effect of workers' compensation insurance on workplace safety.

Greater Experience Rating of WC Premiums

The unit price of workers' compensation insurance coverage depends on two factors, a manual rate capturing the average workers' compensation claims of the industry and an experience modification factor reflecting the firm's injury experience. What are known as primary losses (high frequency, low severity) influence the experience modification factor of all firms, and what are known as excess losses (low frequency, high severity) affect a firm's experience modification factor via the firm's credibility factor. Large firms have high credibility factors and small firms have low credibility factors. The practical implication of credibility factors is that a large increase in relatively severe injuries still has a small effect on WC premiums in small firms. Increases in severe injuries have substantial effects on the WC insurance prices paid by larger firms. Our baseline simulation have a credibility factor of 0.33, which is a 33 percent degree of experience rating for large losses.

In the 1970s the then National Commission on State Workmen's Compensation Laws argued that insurers should place greater emphasis on a firm's own safety record when pricing WC coverage to encourage greater employer specific responsibility for increased job safety and to shift the economic burden of injuries toward high injury rate employers. As in our other simulations we considered several refinements: increasing the experience rating credibility factor by 10 percent, 20 percent, and 200 percent (which is complete experience rating or self insurance).

Table 5-2 indicates a slight safety enhancing effect of greater experience rating. The elasticity of injuries with respect to experience rating is about -0.02. The improved safety that results under improved experience rating also reduces WC applications and claims paid. Our numerical simulations suggest that experience rating has only a small effect on safety and WC claims. Going to complete experience rating, a 200 percent increase in the degree of experience rating, increases WC claims and safety by 0.5 to 0.75 percent.

Reducing Exaggerated WC Claims Paid

We first investigated the effects of inaccurate state verification by considering several improvements in screening out exaggerated WC claims. Reductions in ψ_1 in our simulation model represent more rejections of exaggerated claims; a 100 percent reduction in ψ_1 (to 0) means that all exaggerated claims are denied. Interestingly, Table 5-3 shows that improved screening accuracy slightly raises work-related injuries. The change in injuries is the result of two opposing forces set in motion via the screening mechanism as it applies to exaggerated claims. Reducing the proportion of exaggerated claims paid raises the cost to a worker from taking a relatively risky job, which tends to promote a rematching of workers with employers with higher safety levels. The opposing force is that when fewer exaggerated claims are paid the WC insurance premium of the typical firm falls, giving less financial incentive for employers to provide workplace safety. The movement of workers toward safer jobs and the offsetting effects on

Table 5-3. The Effects of WC Screening Accuracy on the Probability of Injury, Claims Filed, and Claims Paid [Percentage Changes (Elasticities) from the Baseline Simulation in Figure 5.1 and Table 5–1]

Screening — The Probability of Receiving WC Benefits for an Exaggerated Claim (ψ_1)

Probability of Receiving Benefits (% change)	Injuries	Application Rate — WC Benefits	Application Rate — PTD Benefits	Application Rate — PPD Benefits	Frequency of WC Claims Paid	Frequency of PTD Claims Paid	Frequency of PPD Claims Paid
0.306 (-10%)	0.08 (-0.01)	-1.26 (0.13)	-1.80 (0.18)	* (*)	-1.75 (0.18)	-3.24 (0.32)	-0.26 (0.03)
0.272 (-20%)	0.14 (-0.01)	-1.70 (0.09)	-3.61 (0.18)	2.72 (-0.14)	-2.98 (0.15)	-6.26 (0.31)	0.31 (-0.02)
0 (-100%)	0.18 (*)	-13.13 (0.13)	-37.88 (0.38)	44.43 (-0.44)	-11.09 (0.11)	-22.80 (0.23)	0.64 (-0.01)

Table 5-3. *(Continued)*

Screening — The Probability of Receiving WC Benefits for a Genuine Claim (ψ_2)

Probability of Receiving Benefits (% change)	Injuries	Application Rate — WC Benefits	Application Rate — PTD Benefits	Application Rate — PPD Benefits	Frequency of WC Claims Paid	Frequency of PTD Claims Paid	Frequency of PPD Claims Paid
0.77 (10%)	0.03 (0.003)	0.03 (0.003)	-0.77 (-0.08)	1.90 (0.19)	5.48 (0.54)	7.26 (0.73)	3.70 (0.37)
0.84 (20%)	0.09 (0.004)	0.10 (0.005)	-2.45 (-0.12)	6.04 (0.30)	10.45 (0.52)	13.98 (0.70)	6.90 (0.35)
1.0 (42.86%)	0.16 (0.004)	0.14 (0.003)	-2.45 (-0.06)	6.18 (0.14)	19.95 (0.47)	31.66 (0.74)	8.22 (0.19)

Note: Table 5-1 lists the starting parameter values used in the above simulations. All frequencies, probabilities, and rates are relative to the total labor force.

*Less than 0.005

firms, which reduces firms' safety expenditures, creates the result that when WC claims are screened to eliminate a greater number of exaggerated claims there is a slight increase in the average injury level.

Improvements in the ability to separate exaggerated from genuine claims reduces applications for benefits even though injuries increase slightly. The elasticity of the WC claim application rate with respect to the screening accuracy parameter, ψ_1, is about -0.1. Fewer workers file for WC benefits because there is a relatively lower payoff from exaggerating a nonimpairing injury into a compensable injury as well as less incentive to exaggerate a permanent partial disability into a permanent total disability. The elasticity of applications for PTD benefits with respect to ψ_1 is about -0.20 to -0.40 in our numerical simulations. The incentive not to exaggerate PPDs into PTDs means that the application rate for PPD benefits increases as the probability of paying an exaggerated PTD claim drops.

Improved screening of exaggerated claims reduces total WC claims paid because it leads to fewer applications for WC benefits as the expected payoff from an exaggerated claim drops. A 10 percent reduction in ψ_1, the probability a worker receives benefits for an exaggerated claim, lowers total WC claims paid by about two percent and PTD claims paid by two to three percent. Improvements in screening exaggerated claims increase the number of PPD claims paid because the distribution of claims filed moves away from PTDs toward PPDs.

As in simulating WC benefit effects, studying improvements in eliminating exaggerated claims illustrates the care needed in interpreting the safety impact of public policy. A 20 percent reduction in ψ_1 reduces WC applications by about two percent, reduces WC claims paid by about three percent, but increases the injury rate by 0.1 percent. From our simulations one would conclude that improved state verification reduces injuries as measured by WC applications or claims paid, but would conclude that improved state verification raised injuries slightly as measured by actual injuries.

Increasing Genuine WC Claims Paid

The other dimension of imperfect state verification in our simulation model is the probability that a worker receives payment for a genuine WC claim. Improved state verification takes the form of an increase in the state verification parameter, ψ_2, the likelihood of receiving WC benefits for a genuine work-related injury. The improvements in ψ_2 that we simulate and tabulate in Table 5–3 include 10 and 20 percent improvements as well as the elimination of imperfect state verification ($\psi_2=1.0$). The economic effect of raising ψ_2 is a shift of the burden of serious injuries away from workers toward firms.

Increasing the number of genuine claims paid under WC slightly increases workplace injuries. Specifically, the elasticity of the injury rate with respect to ψ_2 is about 0.004. Our simulations reveal that the shift in accident costs away from workers toward firms as a higher proportion of genuine WC claims are paid is relatively small, which accounts for the small increase in worker injuries.

The application rate response to improved screening accuracy via increasing ψ_2 is also small. In contrast to the economic effects of lowering the probability of paying an exaggerated claim (ψ_1) increasing the probability of paying a genuine claim (ψ_2) does not influence the number of NII workers who apply for WC benefits. Because it raises the opportunity cost of filing an exaggerated claim, increasing ψ_2 leads to fewer exaggerated PTD and more genuine PPD claims. In our simulations the elasticity of PTD applications with respect to ψ_2 is about -0.1 and the elasticity of PPD applications with respect to ψ_2 is about 0.1 to 0.3.

Although increasing the screening accuracy of WC via paying a higher proportion of genuine claims has modest effects on injuries and applications for WC benefits the number of claims paid rises substantially. A 20 percent increase in the likelihood of paying a genuine claim raises total claims paid by about ten percent, PPD claims paid by about 14 percent, and PPD claims paid by about seven percent. Claims paid rise with screening accuracy, ψ_2, largely because of workers' greater ability to demonstrate a genuine injury not because of a reduction in overall workplace safety.

Sensitivity of Simulated Outcomes to Structural Parameter Values

A focal point of our book is that numerical simulation of structural models is a useful complement to econometric research. One of the ways we demonstrate the complementarity between estimation and simulation is to use our simulation model to locate where more precise parameter estimates would change the economic and policy implications versus where parameter values matter little to outcomes of interest. We now describe, as in previous chapters, results of how some key structural parameters and starting values affect the economic implications of workers' compensation insurance for job safety, claims filed, and claims paid.

The production function is one of the most researched structural equations in economics, and the first place we examine the sensitivity of our simulation results to numerical background information concerns the degree of input substitutability in output production. Our base case in the results tabulated in Tables 5-2 and 5-3 is a Cobb-Douglas type production function where the partial elasticities of substitution in production (σ) are equal to 1.0 for all three input pairs. We changed σ to 0.90, and recomputed hedonic labor market equilibrium. We also changed the starting values of relative injury incidence to see how increasing f_1, the number of NIIs, to twice the number of PPDs and to see how increasing f_2, the number of PPDs, to twice the number of PTDs affect hedonic equilibrium. The results of our three structural parameter sensitivity checks are that the degree of substitutability in production and the relative injury mix has little effect on the probability of injury or on the number of applications and WC claims paid.

The results in Tables 5-2 and 5-3 also reveal the implications of extreme changes in the basic parameters of the workers' compensation system such as the degree of experience rating and the accuracy of state verification. Applications for claims are the most sensitive to extreme changes in the insurance program parameters with claims paid moderately sensitive, and injuries insensitive to either experience rating or accuracy of state verification.

CONCLUSION

Workers' compensation insurance is the most important public policy influence on workplace safety in the United States. Here we have examined more deeply the quantitative properties of the WC program, specifically the imperfect verification of injuries as work related or their severity. Our numerical simulations provide quantitative insights into an issue currently at the forefront of theoretical microeconomics — information asymmetries (Hirschleifer and Riley, 1992). The practical implication of information asymmetry in the case of workers' compensation insurance is that increases in benefit generosity can paint a puzzling picture of labor market outcomes because the number of workers applying for and receiving WC benefits can rise while actual workplace injuries fall. Data on injuries reported under workers' compensation can produce a wrong conclusion of the effects of WC on workplace safety.

NOTES

[1]Permanent partial and temporary total disabilities account for about two-thirds of total workers' compensation insurance benefits and about one-fifth to one-fourth of claims filed. We simplify by ignoring temporary total disability injuries.

[2]In the expected utility function in (5.1) differences in α across workers totally parameterize interpersonal heterogeneity. The Arrow-Pratt measures of relative and absolute risk aversion are directly related to α, and differences across workers in α represent the heterogeneity in attitudes regarding injury risk and willingness to apply for (stochastic) workers' compensation insurance benefits. Risk aversion requires $0<\alpha<1$. More importantly, our assumptions concerning a single parameter capturing worker heterogeneity mean that we are abstracting from interpersonal differences in human capital, WC coverage, the probability of receiving WC benefits, the disutility of a disabling injury, the disutility of receiving benefits from an exaggerated WC claim, available leisure time, the value of leisure, and entitlement to benefits from public or private income maintenance programs.

[3]Safety measures can increase production by reducing interruptions due to injuries and by increasing the continuity of employee attendance (Viscusi

1979). However, some safety measures, such as slowing the pace of the assembly line or introducing cumbersome machine guards to reduce injuries, decrease output. Whether safety expenditures are on balance production enhancing is an unresolved empirical question. So not to downplay the role of government regulations requiring safety enhancing measures we again took safety as contributing positively to production.

[4]Workers who receive total disability payments or who are seriously injured provide no work effort. Moderately injured workers who do not apply for WC benefits contribute $d(<1)$ units of labor and workers denied either PPD or PTD benefits contribute $\tilde{d}\,(<d)$ units of labor. For more details see Kniesner and Leeth (1989b, footnote 4).

[5]Worker heterogeneity appears via interpersonal (stochastic) differences in the risk aversion parameter (α), which also determines whether a worker files an exaggerated WC claim. The condition for labor market equilibrium in (5.5) also has as an outcome that each firm, which is uniquely represented by a value of safety efficiency (μ), hires a homogenous work force (as captured by a common value for α). The result of the matching mechanism in our simulation model is that at some levels of safety (μ) all workers with nonimpairing injuries apply for WC benefits ($a_1=1$) and all PPD workers apply for PTD benefits ($a_2=1$).

[6]In solving numerically for hedonic labor market equilibrium in (5.5) we respecified the differential equations (5.6)–(5.8) into three difference equations that locate α, μ, and w at 13 pre-chosen probabilities of injury. We then jointly solved the nonlinear difference equations representing (5.5)–(5.8) via Newton's method, which is a gradient or hill climbing procedure. Chapter 3 has more details of the numerical simulation method we used.

[7]We emphasize that workers' compensation insurance differs from social insurance under Aid to Families with Dependent Children or Social Security Disability Insurance, which are available as income support to persons with generally low incomes. Injured workers receive no-fault insurance coverage in exchange for the inability to sue their employers under tort laws.

[8]We computed the second-order conditions for workers and employers to verify that our initial w–π locus represented a stable equilibrium. In all cases the second-order conditions for worker utility maximization and employer profit maximization were satisfied.

6 POLICY IMPLICATIONS

To now we have focused on how labor markets in the United States and elsewhere establish implicit payments, known as compensating wage differentials, for work-related injury risks (Kniesner and Leeth 1991b). Even though the labor market lets firms lower wages if they make their workplaces safer, there has been no decline over time in work-related injuries, which has been interpreted by policymakers as insufficient market generated incentives for workplace safety. The two pillars of job safety policy in the United States are the states' workers' compensation insurance systems (WC) and the federal government's Occupational Safety and Health Act of 1970 (OSHA). Here we discuss recent changes in U.S. workplace safety policy and suggest cost effective reforms to OSHA and WC.

Our policy recommendations begin with the overwhelming evidence that attempting to reduce workplace hazards via OSHA's regulatory approach has not been cost effective. OSHA should no longer set standards for the organization of jobs and production, and the agency's budget should be folded into the National Institute of Occupational Safety and Health (NIOSH). Workers' compensation insurance can be more completely experience rated to give employers increased incentives to take safety enhancing measures. Waiting periods before receiving WC benefits could also be increased to encourage greater safety efforts by workers. We conclude by discussing the possibility of changing workers' compensation insurance to place more emphasis on determining fault.

OSHA DURING THE 1970S VERSUS THE 1980S[1]

The Occupational Safety and Health Act of 1970 established mandatory workplace characteristics backed by inspections and fines for noncompliance with OSHA's regulations. Under OSHA workers also cannot be expected to use information on job related health hazards to take injury avoidance measures on their own or to use personal safety devices if provided by the firm. Standards in turn require plantwide capital investments. The logic of mandatory health and safety standards for employers is that a government agency specializing in workplace safety will know more than employers and workers concerning how to make work sites safer.

Firms' opposition to OSHA centered on the belief that the regulations would not produce cost effective improvements in workplace safety. The benzene and cotton dust cases established the principle that OSHA must be cost conscious only when a standard is either totally ineffective, technologically impossible, or so costly for firms that the industry would no longer be economically viable.[2] Our discussion of OSHA during the Nixon, Ford, and Carter administrations of the 1970s and the Reagan and Bush administrations of the 1980s focuses on the continuing debate over whether OSHA regulations should be subjected to cost effectiveness checks.

OSHA During the 1970s

The Occupational Safety and Health Administration initially issued over 4,000 safety standards that included the design and placement of exit signs, specification of handrails, and the requirement of portable toilets for wranglers. During the 1970s OSHA's regulation activities grew rapidly as metered by its budget, personnel, and the number of regulations OSHA issued. By 1980 the Occupational Safety and Health Administration employed about 3,000 people and had a budget of about $178 million. OSHA spent a total of about $1.4 billion (1982 dollars) during the 1970s to establish and enforce workplace safety standards including widely debated maximum exposure levels to arsenic, benzene, cotton dust, lead, and noise. The initial goal of the backers of OSHA was to cut workplace injuries in half (Viscusi 1992, p. 248).

There is convincing evidence that OSHA's standards were ineffective in raising workplace safety during the 1970s (Smith 1992). Complying with the initial standards under OSHA would have cost the average firm at least $300 per worker. If a firm did not comply with the standards in place during the 1970s its expected fine was only about $17 per worker, however. The low expected fines were the result of both low inspection probabilities and low fines if discovered out of compliance. The typical firm's economically rational behavior during the 1970s was not to comply with OSHA standards until caught.

The standard setting approach to enhancing workplace safety also is inherently ineffective. Estimates from California, New York and Wisconsin showed that about only 10–25 percent of injuries resulted from violation of a safety standard. Another finding was that about only 25–36 percent of injuries were caused by hazards controllable by safety standards. Using the midpoints of the ranges of how safety standards control injuries and how safety standards are obeyed suggests that the maximum effect of OSHA would be a 25 percent reduction of workplace injuries (Smith 1979, 1992).

Numerous econometrically sophisticated papers studying the impact of OSHA during the 1970s while controlling for many intervening factors, including worker quality and the pace of production, have come to a common conclusion. There is no evidence that OSHA had *any* safety enhancing impact during its first decade (Smith 1992). OSHA was a program with costs and no accompanying safety benefits.

As a final thought we reiterate that Congress and the courts have established that the Occupational Safety and Health Administration need only be cost conscious when issuing a new workplace safety standard if the standard is technologically impossible, so expensive that conforming with the standard will put the industry out of business, or if the standard does not improve worker health or safety. During the 1970s OSHA's collection of activities seemed to be ineffective at their stated goal, which is to enhance workplace safety in the United States. By failing to have any safety enhancing effect the Occupational Safety and Health Administration of the 1970s flunked the test of cost consciousness that individual OSHA workplace safety standards must pass.

OSHA During the Reagan Administration

The Reagan administration tried to create a cost effective OSHA. The agency's budget and staff shrank during the Reagan years. Employment in the Occupational Safety and Health Administration fell about 20 percent, and OSHA's real annual budget ($1982) fell steadily from a high of $206 million in 1981 to about $180 in 1987. There were two new major standards issued during President Reagan's first term compared to eight major standards during the one term Carter presidency. The most important standard issued by OSHA during the Reagan years was a set of rules intended to reduce grain handling accidents, where the employer was permitted to choose from a set of options for decreasing grain dust. The Reagan OSHA formulated new standards more slowly, and new OSHA standards also gave firms some flexibility concerning what would satisfy the regulations.

The Reagan OSHA paid increased attention to health, as opposed to accident, standards. Citations for violations of health related standards were about 16 percent of all OSHA violations during 1983 compared to about two percent in 1976. The Reagan administration also redirected inspections away from manufacturing and toward construction, the most dangerous industry as measured by industrial accident rates. During 1983 approximately 50 percent of federal OSHA inspections were of construction sites compared to 25 percent of all OSHA inspections encompassing construction sites during a typical year of the Carter OSHA. During 1981-88 more of OSHA's efforts were placed into areas where regulation could have the greatest safety enhancing effect: health hazards and comparatively dangerous work sites.

In 1981 as an attempt to get more injury deterrence for its money OSHA began to use firms' safety records to target inspections at work sites with comparatively poor safety records. OSHA began to inspect only firms with lost workday accident rates for the past two to three years that were above the most recent overall manufacturing accident rate. Using firms' injury records to select sites for inspections was eventually discontinued partly because of concern that government injury data were becoming less accurate over time as firms under reported injuries to avoid inspections.

Because records checks were used instead of site visits in about 15 percent of inspections and because a construction site inspection counted each subcontractor as a separate inspection the total number of workers visited by OSHA inspectors dropped by over 40 percent during the Reagan administration. During the first Reagan term, 1981-1984, OSHA inspections covered only about two to three million workers annually.

The level and motivation behind fines for OSHA violations also changed during the 1980s. Total OSHA fines during 1983, for example, were about $6 million compared to $26 million of OSHA fines during the last year of the Carter administration, 1980. The Occupational Safety and Health Administration also became less confrontational and more cooperative with firms concerning safety matters. One result of less adversarial safety regulation was that firms contested relatively fewer OSHA fines. Firms contested three percent of fines in 1983 versus 21 percent in 1979. Although OSHA imposed fewer penalties a greater proportion of penalties assessed were for serious violations — 70 to 80 percent of penalties during 1982–83 were for serious violations compared to 50 to 60 percent during the late 1970s. Overall, the Reagan administration's OSHA issued fewer citations and lowered total fines, and firms contested relatively fewer fines during the 1980s.

OSHA During the Bush Administration

The Bush administration continued to redirect inspections and new regulations away from the stereotypical manufacturing workplace and toward more hazardous industrial settings, that included logging, construction, agriculture, maritime, and petrochemicals. The Bush administration's OSHA also began to consider worker protection against cumulative trauma disorders and blood borne diseases. Legislation during the Bush administration also increased by a factor of seven the maximum penalties a firm could receive for noncompliance, to $70,000. Total OSHA fines collected during fiscal year 1990 were about $30 million, the largest in OSHA history.

OSHA during the Bush administration encouraged greater use of seat belts while driving at work. Moreover, as many as 120 deaths and 28,000 serious injuries happen annually in manufacturing because a worker accidentally switches on a machine that another worker is servicing. The Bush OSHA issued regulations requiring a worker doing machine maintenance either to lock or tag the machine so that another worker either cannot or will not try to use the machine undergoing maintenance. Finally, it has been estimated that 800–1600 workers die each year from diseases related to passive smoking at work. No chemical substance related to as many deaths as passive smoking would have continued to escape regulation, and the Bush administration's OSHA sought reductions in passive smoking at work. Under President Bush OSHA emphasized cheap, but effective, safety enhancing regulations.

An Evaluation of OSHA During the 1980s

Unlike evidence for the 1970s some signs of an OSHA effect on workplace accidents appear in the econometric studies encompassing the 1980s. Evidence including the mid-1980s suggests OSHA inspections reduced lost workdays by two to seven percent (Smith 1992). The estimated safety enhancing impact of OSHA during the 1980s is still small, however, and in some research the result of having enough years of data to attain statistical significance. During the 1980s OSHA still had a low inspection rate (0.5 percent of firms inspected per year was typical), found few violations found per inspection (typically two), and imposed low fines per OSHA violation discovered (typically \$60 per violation or about \$0.60 per worker). In our judgment OSHA continued to have an inconsequential influence on workplace safety in the United States.

Financial incentives from other sources overshadow OSHA's meager economic incentives for greater workplace safety. The \$6 million of OSHA fines during 1982 are tiny compared to the approximately \$10 billion firms paid in workers' compensation insurance premiums and the estimated \$69 billion in compensating wage differentials firms implicitly paid for having less that completely safe workplaces. Estimates are that investments of \$100 to \$500 billion was required for firms to comply fully with OSHA regulations proposed during 1975–1980 alone. A back-of-the-envelope calculation using the range of estimates of OSHA's effectiveness in reducing injuries during the 1980s indicates that complying with OSHA regulations would have cost firms about five times more than the safety benefits.[3] A program as cost ineffective as OSHA needs major reform.

Recent Developments in OSHA

New safety standards are being considered that reflect recent health concerns. Specifically, OSHA has proposed ways to prevent blood borne diseases (*Occupational Hazards* 1991), introduce ergonomics in injury prevention (*Occupational Hazards* 1993b), limit indoor smoking at the workplace (*Time* 1994), regulate work done in confined spaces (*Daily Labor Report* 1992b), and establish permissible limits for exposure to glycol ethers, methylene chloride, and 1,3-butadiene (*Daily Labor Report* 1994b).

OSHA's direction in the area of enforcement reflect the Clinton administration's announced tough attitude toward lawbreakers plus their intention to reinvent government. In particular, OSHA has been reevaluating its severity of civil penalties (*Daily Labor Report* 1994a) and criminal penalties (*Daily Labor Report* 1992a) for noncompliance.

Vice-President Gore's National Performance Review of Government proposed that employers either hire inspectors from private inspection firms or have employees trained and certified to conduct inspections (Salwin 1993). OSHA would set private inspection standards and conduct random reviews of the private inspections. The suggestion for self inspection is in response to the Gore committee's conclusion that OSHA is unable to inspect and enforce standards adequately without impossibly high increases in funding (*Occupational Hazards* 1993a).

The direction that OSHA has taken during the 1990s falls into two categories: introducing new or revising existing standards to try to improve workplace safety and finding new or revising old ways of enforcing OSHA's existing standards. Having summarized what OSHA has been up to lately we now discuss possible reforms that are come from our interpretation of the evidence both in econometric studies summarized in Chapter 2 and the numerical simulations in Chapters 4 and 5.

REFORMING OSHA

From our review of the econometric research and our simulations in previous chapters we conclude that OSHA has been cost ineffective because it has been economically irrelevant to firms' decisions concerning workplace safety. OSHA's meager economic incentives, which are due to its low expected fines for noncompliance and high compliance costs to firms, lead to widespread noncompliance with OSHA regulations. A cost effective Occupational Safety and Health Administration would stop setting workplace safety standards at such severe risk levels.

Some economists argue that the law underlying OSHA could be changed to make current and proposed new OSHA standards pass cost effectiveness tests. Standards would need to take firm heterogeneity into account, so that the most stringent (costly) standards would be placed on firms whose compliance costs are the lowest. Not only, then, would different standards be applied to the same substance across industries but different safety standards would also be applied to different stages of production within a firm. In our judgment requiring new and current standards to pass cost effectiveness checks would lead to a protracted political battle over what to include in the costs and benefits of each proposed regulation with little eventual reform in how OSHA operates.

Another possible reform to OSHA would be to redirect the agency away from formulating workplace standards and move OSHA toward setting target safety levels. Making OSHA focus on safety levels as opposed to workplace safety standards would be similar to current environmental protection policy where the focus is on emission levels rather than on how firms reduce pollution. Economic incentives emphasizing improved safety already exist in workers' compensation insurance, and having OSHA fine firms for failing to meet an injury rate target would be redundant.

In our judgment OSHA has been a public policy failure from an economic efficiency or cost effectiveness viewpoint. We recommend that the Occupational Safety and Health Administration no longer issue mandatory workplace standards, inspect firms for compliance with OSHA standards, or fine firms for noncompliance with safety standards. OSHA's budget should be folded into the National Institute of Occupational Safety and Health (NIOSH). We believe that resources now used to regulate the workplace should be redirected to researching what workers and firms can do to reduce injuries and diseases at work and to providing the results of NIOSH research to firms and workers.

As evidence of the likely effectiveness of our recommendation that OSHA disappear into an expanded NIOSH we cite a recent study of the response of chemical workers to information concerning possible new workplace health hazards (Viscusi and O'Connor 1984). Workers told that they would soon be working with sodium bicarbonate, a safe chemical, reduced their assessment of workplace hazards by 50 percent. Workers told that they would be working with either asbestos or TNT increased their assessments of workplace hazards by 200 percent. No workers required extra compensation to handle the safe sodium bicarbonate, but workers demanded an extra $3,000 to $5,000 per year to handle the dangerous asbestos or TNT. No workers said they would quit their jobs because they would be handling sodium bicarbonate, but a majority of workers said they would quit because they would be handling asbestos or TNT. The Viscusi and O'Connor study of chemical workers demonstrates that information can influence industrial safety by adjusting workers' risk perceptions and their willingness to work in a situation.

The federal government would better serve the taxpayers at large and workers and firms in comparatively hazardous industries by having OSHA's resources transferred to NIOSH. An expanded National Institute of Occupational Safety and Health could then do more research on the causes of workplace health hazards, especially occupational diseases. The goal of our policy recommendation is to provide more information and guidelines to firms and workers concerning danger thresholds of exposure to toxic substances and dangerous workplace practices.

WORKERS' COMPENSATION INSURANCE AND SAFETY

Workers' compensation insurance (WC) is largely state operated disability insurance covering workers for economic and health losses from industrial accidents and some diseases. Federal programs cover federal employees and longshoring and harbor workers. Federal legislation also gives benefits similar to workers' compensation to workers in coal mining, railroading, and maritime shipping. Regardless of the party at fault employers must fully compensate injured workers for medical expenses and partially compensate workers for earnings foregone. Most states make firms replace two-thirds of injured workers' weekly wages tempered by minimum and maximum amounts. Low wage injured workers will collect minimum benefits, which will be more than two-thirds their pre-injury wages, and high wage injured workers will collect maximum benefits, which will be less than two-thirds their pre-injury wages. States generally permit firms to offer workers' compensation insurance either by purchasing coverage from a private provider or by self insuring after proving an ability to pay claims. In return for the no fault insurance coverage under WC workers cannot sue employers for damages due to work-related injuries. Because the insurance premiums that firms pay for WC reflect the level of injuries workers' compensation insurance provides economic incentives for employers to create safer workplaces.[4]

Before workers' compensation insurance an injured worker had to prove employer negligence to collect damages. An employer could avoid negligence damages by demonstrating in court that the worker's injury was the worker's fault, was an ordinary hazard of employment, or was caused by a coworker's carelessness. Under the pre-WC fault determination system compensation for injuries was uncertain. Many viewed the negligence system as providing insufficient protection for workplace injury risks. The pre-WC fault system with its uncertainty and escape clauses for employers may also have provided little economic incentive for employers to improve workplace safety. In contrast to the pre-WC common law setting a workers' compensation insurance system provides economic certainty for injured workers and a stronger economic incentive for employers to improve workplace safety.

Using total expenditures as a measure of size the WC program is now larger than food stamps, Aid to Families with Dependent Children, and unemployment insurance. In 1987 WC covered about 88 million workers, 87 percent of the labor force, and paid about $25 billion in total benefits, approximately 1.4 percent of total payroll. Including administrative expenses the total cost of workers' compensation insurance is now about two percent of total U.S. payroll expenses (Schwartz, et al. 1989).

Recent Developments in Workers' Compensation

Recent state legislation has made detailed revisions to standing policy. The most commonly occurring change in states' workers' compensation laws during the first half of the 1990s has been increasing maximum compensation rates coupled with wider use of deductibles that have the effect of reducing payments for relatively minor injuries (Tinsley 1991; Berreth 1992, 1994; Brown 1993).

The Safety Incentives in Workers' Compensation

By 1970 many policy makers argued that the workers' compensation system was not keeping its initial promises to encourage workplace safety and to protect injured workers against income losses. From 1958 to 1979 the average manufacturing injury rate grew by about two percent per year (Viscusi 1983). Because most states changed their maximum benefits infrequently WC benefits fell in real terms during the 1960s. WC benefit maximums were capped at relatively low levels while wages rose during the 1960s. Few disabled workers received the recommended 67 percent of pre-injury wages as workers' compensation benefits during the 1960s.

The upward trend in injury rates during the 1960s helped motivate the Occupational Safety and Health Act of 1970, which recommended creating the so-called National Commission on State Workmen's Compensation Laws to evaluate the insurance adequacy and safety incentives in WC. The National Commission recommended that states' WC laws provide benefits for temporary total disabilities, permanent total disabilities, and deaths that were the smaller of 67 percent of the worker's pre-injury wage and 67 percent of the state's average weekly wage. The National Commission's recommended reforms to WC were taken seriously because the Commission recommended states be given three years to comply after which Congress should pass legislation forcing states to comply with the Commission's recommendations (Berkowitz and Berkowitz 1985).

During the 1970s the threat of federalization led to a rise in state WC benefits. Of the 50 states, Puerto Rico, the Virgin Islands, and the District of Columbia in 1972, only six had the recommended maximum income benefits for temporary total disabilities of two-thirds the state's average weekly wages. By 1983 at least 32 of the 53 U.S. jurisdictions had met National Commission standards for benefits for temporary total disabilities, permanent total disabilities, and deaths (Berkowitz and Berkowitz 1985). Increased compliance with the National Commission's recommendations also made income compensation rise by 725 percent during 1970–1986 to $16.5 billion. Accompanying the rise in benefit generosity was a rise in WC premium costs. During 1970–86 workers' compensation premiums doubled from about one to about two percent of payroll of covered employees (Schwartz, et al. 1989). The doubling over time in WC benefits plus differences in benefit levels across states allowed researchers to identify some of the safety effects of WC benefit generosity.

Econometric Evidence. A typical result from econometric research is that higher replacement of lost earnings raised the number of workplace injuries reported (Moore and Viscusi 1990). Estimates are that during the 1970s increased workers' compensation benefits caused about half the increase in temporary total and minor permanent partial disability injuries and about one-third the increase in the number of major permanent partial disabilities reported (Butler and Worrall 1983). The estimated elasticity of reported injuries to benefit generosity is about 0.4, or a ten percent increase in workers' compensation benefits leads to about a four percent rise in work-related injuries reported.

Researchers have also examined how WC benefit generosity influences injury duration. One study of the duration of lower back injuries, chosen because their existence and severity is difficult to verify medically, found an elasticity of injury duration with respect to benefit generosity of 0.4. Ten percent greater WC benefits raises the duration of lower back related temporary total disability claims by about four percent (Butler and Worrall 1985, Krueger 1990b). Another study examining the average recovery of all WC cases in Minnesota found that a five percent increase in workers' compensation benefits raised the time away from work by about eight percent (Krueger 1990b). The higher are WC benefits the lower is the incentive for an injured worker to return to work, and econometric research verifies a positive relation between WC benefits and the length of time before an injured worker returns to employment.

One of the goals of the workers' compensation insurance system is to encourage greater workplace safety. The econometric research we have been describing suggests a conflict between the income security that WC provides injured workers and the incentives that WC provides workers and firms to avoid injuries. Higher WC benefits raise employers' insurance premiums, which can be offset by making the workplace safer so that workers file fewer WC claims. WC pricing encourages firms to make greater investments in safety enhancing measures. Higher workers' compensation benefits reduce workers' losses from injuries and may discourage individual safety efforts or lengthen recovery periods, however. Because safety incentives on employers and workers conflict, increasing WC benefits will increase the overall frequency and duration of work-related injuries if workers' incentives via WC dominate firms' incentives via WC.

Actual Versus Reported Workplace Injuries. Although workers may exercise less care to avoid injuries as WC benefits grow workers are also more likely to report an injury and file a WC claim. In our judgment the positive relation between WC benefits and injuries is more the result of increased injury reporting than a decline in workplace safety. For the typical worker the most important part of a work-related accident or disease is not the lost income but rather the pain and suffering. Econometric estimates are that to avoid a single fatal accident workers would collectively sacrifice $2 million to $6 million in wages (Moore and Viscusi 1990). Because workers place such a high collective value on a life it is difficult for us to conclude that the comparatively small increases in workers' compensation benefits in the United States since 1970 have increased actual injuries by much.

Research suggests that incentives to file an accident claim produce serious wrong impressions of how workers' compensation insurance influences industrial safety. Insurance providers cannot determine exactly the severity of an injury or whether the injury was work-related. Because of the information asymmetry between workers and insurers some workers who have legitimate WC claims will be denied benefits, and some workers who have nonimpairing injuries or injuries unrelated to work will receive workers' compensation benefits. Higher WC benefits give an incentive for workers with spurious claims to report injuries and to file for workers' compensation benefits.[5]

Because work-related fatalities or serious injuries are hard to dispute workers or their families have little economic incentive to file spurious claims. Serious injuries should be relatively free of the reporting effect and can be used to isolate the true influence of WC on workplace safety. Between 1900 and 1940 the introduction of a WC law was followed by a reduction in the state's industrial death rate, ceteris paribus (Chelius 1977). Another recent study found that a ten percent increase in WC benefits lowered occupational fatalities by about four percent (Moore and Viscusi 1990). Higher WC benefits reduce fatal or severe injuries.

In the previous chapter we numerically simulated employer and worker safety incentives and workers' decisions to file WC claims. In our simulations doubling the WC income replacement rate from its recent historical level would reduce the injury rate by two percent while raising reported injuries by about 30 percent. Our numerical simulations illustrated that when WC benefits increase the actual rate of work-related injuries can fall while the application rate for workers' compensation benefits rises.

Our simulations and other researchers' econometric results underscore workers' compensation insurance provides economic incentives for employers to make their workplaces safer. The positive influence of WC on safety appears when actual injuries are measured instead of injury rates that reflect reporting bias. Public policy must cope with the fact that higher WC benefits lead some workers to file spurious claims for compensation and to spend more time away from work recovering from compensable injuries.

REFORMING WORKERS' COMPENSATION

We now discuss four frequently mentioned changes in states' workers' compensation programs. The four policy changes we examined are raising WC benefits, expanding the degree of firms' experience rating, allowing workers to sue employers for negligence, and instituting a dual insurance system mimicking the U.S. automobile insurance system. Our focus is on determining whether any of the four possible policy changes in WC could increase worker safety without unreasonably increasing employer costs.

Raising Workers' Compensation Benefits

The most important force constraining the level of WC benefits is the growing cost of coverage to employers. Since 1970 the dollar costs of medical and income support have risen by over 700 percent, from about one percent of payroll to about two percent of total payroll (Schwartz, et al. 1989). Because they are responsible for paying WC insurance premiums directly, employers interpret the financial burden of WC as totally theirs and have been a vocal opposition to increasing WC benefits. In labor market equilibrium wages fall as WC benefits rise, ceteris paribus, so that workers implicitly pay a share of the cost of workers' compensation insurance.[6] Workers overall, who indirectly pay part of the insurance premiums, may also oppose higher WC benefits for injured workers. Econometric research indicates that workers do not now want and are unwilling to pay for further increases in WC benefits (Moore and Viscusi 1990, Viscusi and Evans 1990). Neither firms nor workers appear to have much interest in expanding workers' compensation payments and premium costs, and we too oppose a general increase in WC benefit generosity as a policy to reduce industrial injuries.

A policy alternative that deserves consideration is to lengthen the waiting period before an injured worker receives WC benefits. The longer a state requires a worker to be off the job before receiving WC the lower the state's injury rate, ceteris paribus (Chelius 1982, Butler and Worrall 1983). A longer waiting period is similar to a tax on workers for low severity, high frequency accidents — accidents that workers may not now be taking cost effective personal actions to avoid. Even if WC replaced all lost income and paid all medical costs workers would still take care to avoid serious injuries because of the pain and suffering involved. For minor injuries the pain and emotional distress may not be enough for workers to be careful to avoid injury. We recommend longer waiting periods before receiving workers' compensation benefits to encourage workers to take more care to avoid low severity, high frequency work-related injuries.

Greater Experience Rating of WC Premiums

As we noted in the previous chapter the price of workers' compensation insurance reflects a manual rate capturing the average WC claims for the industry and an experience modification factor capturing the individual firm's safety record. Primary losses, from high frequency but low severity injuries, affect the experience modification factor of all firms. Excess losses, from low frequency but high severity injuries, affect a firm's experience modification factor depending on the firm's insurance credibility. Firms with few employees have low credibility factors so that increased injuries have little effect on the insurance premiums that small firms pay. Firms with many employees have high credibility factors so that insurance prices rise significantly if injuries increase in large firms. Workers' compensation helps to improve safety when it acts as a tax on employers for the number and severity of work-related injuries.

A typical relationship in states' workers' compensation insurance programs is that a 50 percent reduction in WC injury costs produces no change in WC costs for a firm with seven workers, a five percent reduction in the WC costs for a firm with 10 workers, a 17 percent reduction in the WC costs for a firm with 75 workers, a 36 percent decline for a firm with 750 workers, and a 50 percent (dollar-for-dollar) decline in WC costs for a firm with 1750 or more workers (Chelius and Smith, 1987). Premium changes also take four years to take full effect. Because small firms' insurance prices do not fall much if they improve their safety small firms' safety incentives under WC are limited, and small firms have little reason to contest the severity or genuineness of their workers' WC claims.

Studies have found that when WC benefits increase small incompletely experience rated firms have the largest overall increase in worker injuries and the smallest reduction in work-related fatalities. Moreover, in firms that self insure under WC, which creates complete experience rating, the injured return to work in ten percent less time than average. An increase in WC benefits does not increase the average recovery time of workers in self insured firms but increases injured workers' average recovery time in firms purchasing less than completely experience rated insurance privately (Krueger 1990b). Experience rating matters both for injury rates and average injury duration.

In addition to providing an economic incentive for firms to improve workplace safety experience rating can reduce spurious claims. Giving more weight to injury severity in the experience modification formula makes firms monitor the recovery progress of their injured workers more carefully to minimize the ultimate increase in their WC premiums. WC providers currently view the duration of work-related injuries as beyond employers' control and discount claims of more than $2,000 in the experience rating formula (Krueger 1990b). The low weight for the total cost of a WC claim in future premiums helps explain why firms that switch to self insurance evaluate claims more carefully than firms with the highest experience rating factor in privately purchased insurance. Self insured firms require more second opinions from physicians on the extent of an injury and challenge whether the injury was work-related more frequently (Chelius and Kavanaugh 1988).

Insurance prices could more closely reflect the level and severity of injuries in both small and large firms to encourage greater workplace safety. Environmental economists have argued for effluent charges or pollution taxes to make firms consider the costs and benefits of various pollution control options. By similar logic we prefer injury taxes to health and safety regulations such as OSHA. Firms faced with an injury tax will install additional safety equipment or have a safety education program if the benefits of the expenditure, the taxes avoided via fewer workplace accidents, exceed the expenditure on enhanced safety. Experience rating is a type of tax on workplace injuries encouraging firms to improve safety when economically beneficial. In light of the econometric evidence discussed here and the numerical simulations of the previous chapter we support greater experience rating while noting that increased experience rating alone will not lower injuries by much.[7] For the same reason, we oppose suggestions to consolidate WC insurance into a national health care plan. Consolidating WC into a national health insurance plan would seriously undermine experience rating for workplace safety and health hazards and thereby lower the financial incentives firms now have to provide safe working environments (Countryman 1994).

A simple way to effect a closer relationship between a firm's injury costs and its WC expenditures, the essence of greater experience rating, is to change the typical WC policy to have a larger deductible for an employer (Smith 1992). If an employer must make a sizable copayment to an injured covered worker then the typical firm would bear the financial responsible responsibility for most smaller injuries. A scheme requiring the firm to compensate a worker directly for the first $3,500 of an injury claim would place the financial responsibility for 90 percent of injuries directly on the employer leaving inside the WC system the 10 percent of injury claims that produce 85 percent of total workers' compensation claims expenses (Smith 1992). Greater copayments by firms under WC would leave workers' benefits unchanged while achieving injury rate reductions similar to greater experience rating.

Allow Workers to Sue Employers for Negligence

Workers' rights groups argued that the pre-WC common law system of fault and negligence provided inadequate and unfair compensation for injured workers and few incentives for employers to improve workplace safety. The framers of WC operated from the premise that workers would improve their general level of well-being by sacrificing common law rights to sue employers for negligence in exchange for guaranteed compensation, regardless of fault. Recently researchers have offered an alternative viewpoint to the traditional arguments for the advent of workers' compensation insurance as stemming from meager financial protection of workers under the common law. The alternative explanation for WC is that courts became increasingly willing to find in favor of the employee during the early 1900s, and employers also wanted a no fault workers' compensation insurance scheme to limit their perceived increasing liability for workplace injuries (Chelius 1977).

Allowing workers to sue their employers for negligence while receiving WC benefits would increase employers' expected costs from injuries and encourage greater employer initiated safety efforts. With the no fault system now in effect employers are liable only for a portion of the total costs of injuries. Workers receive compensation for medical expenses and income losses, but no compensation for their pain and suffering. Because employers will eliminate workplace hazards if the expected saving from fewer injuries exceed the necessary financial outlays, employers spend too little for safety under the current limited liability scheme. Permitting workers to sue for pain and suffering would produce greater safety by forcing firms to bear more of the total costs of injuries in situations where wages inadequately reflect uncompensated losses ex ante.

Along the lines of our recommendation the Indiana Supreme Court has recently ruled that Indiana workers injured on the job can sue their employers for civil damages if they can prove their employers intentionally or knowingly harmed them (Associated Press 1994). The Indiana ruling came from a 1991 lawsuit filed by workers who claimed neurological and central nervous system injuries suffered from exposure to PCBs at Westinghouse Electric Corporation plants. More generally, econometric evidence from product liability settlements is that awards for pain and suffering depend positively on injury severity (Viscusi 1988). Allowing workers to sue for damages from pain and suffering should have the effect of deterring injuries. On the negative side, permitting legal actions against employers for work-related injures will increase administrative and court expenses and add uncertainty to compensation for injured workers.

Institute a Dual Insurance System

To a limited degree the current workers' compensation system penalizes nonnegligent parties and rewards negligent parties in an industrial injury. WC makes safety conscious employers pay medical and income replacement benefits when not at fault and only partially compensates careful workers for their injury costs. Similarly, WC less than fully penalizes careless employers by requiring they pay less than full compensation for injuries and less than fully penalizes careless workers by providing support for injuries. Under workers' compensation as currently configured nonnegligent firms and workers will overinvest in safety enhancements and negligent employers and workers will underinvest in safety (Kasper 1975).

A dual insurance system would require both employers and workers to purchase insurance against the costs of work-related injuries. A dual insurance system operates similarly to a strict negligence system while providing the same financial protection for injuries as the current WC system. An injured worker would immediately draw benefits from the employer's insurance carrier. Soon after an accident the two insurance companies would attempt to determine fault, similar to automobile insurance in the United States. If the worker were determined to be at fault then the worker's insurance company would reimburse medical expenses and foregone income. If the employer were at fault then the firm's insurance company would pay medical expenses, lost income, and an amount to compensate for pain and suffering. If under a dual insurance scheme the worker's and the firm's insurance carriers could not agree on fault then the courts would assign fault and damages.

A dual insurance system not only shifts accident costs toward the negligent party but it also encourages greater workplace safety via improving workers' knowledge of safety hazards. Workers now may have little information concerning the hazardousness of a workplace prior to taking a job. Once employed workers learn the risks more accurately and may quit rather than continue working in an environment that is more dangerous than originally believed. Econometric research suggests that wrong perceptions by workers of the true level of workplace accident risk may account for as much as one-third of all quits in the manufacturing (Viscusi 1983). Under a dual insurance system workers would purchase disability insurance prior to employment, and insurance premiums would vary across employers and occupations signaling relative risk differentials across employers. Workers could then refuse to accept a job or demand sufficiently high wages as prior compensation for accepting health hazards at work. The additional information provided by the structure of insurance prices for workers would lead firms to invest in safety in an attempt to hold down the greater implicit wage compensation for injury risk under a dual insurance system.

Many of the desirable aspects of a dual insurance system could be duplicated within the context of the current workers' compensation insurance system by allowing workers to sue for negligence and by lengthening the waiting period before workers begin to receive disability benefits. Allowing workers to sue for negligence shifts the costs of severe industrial injuries toward the firm when it is at fault, which encourages greater safety efforts by careless employers. The uncompensated pain and suffering from a severe injury would still make workers take cost effective safety measures under WC. For less severe injuries a longer waiting period before WC benefits begin would shift accident costs toward negligent workers, and the labor market would establish compensating wage differentials for workplace hazards which would shift accident costs toward negligent employers. The primary advantage of a dual insurance system relative to altering WC to allow negligence suits and to have greater waiting periods is the direct information the worker will receive by interfirm insurance price differentials.

CONCLUSION

OSHA's infrequent inspections and low fines make noncompliance with OSHA standards widespread. Econometric evidence we reviewed shows that OSHA had no effect on safety levels in the United States during the 1970s and reduced lost workday injuries by two to seven percent during the 1980s. A two to seven percent injury reduction is much less than the 50 percent injury reduction the framers of OSHA had intended. OSHA is not cost effective in raising workplace safety levels in the United States, and even unionized workers are now becoming skeptical of the value of OSHA's regulatory approach to safety (Nomani 1994).

As currently configured OSHA does not have the flexibility to achieve cost effectiveness. For example, cost effective safety regulations would place more stringent standards on firms or production processes where compliance costs are lowest. Differential safety standards would apply for the same substance across industries or across stages of production within a firm. We believe that recommending current or new OSHA standards pass cost effectiveness checks would lead only to distracting debates over what to include in the costs and benefits. Instead, we advocate eliminating OSHA and placing its budget in the hands of the National Institute of Occupational Safety and Health (NIOSH). An expanded NIOSH would do additional research into the causes of industrial diseases and provide new information and guidelines to firms and workers concerning unsafe exposure levels for dangerous substances.

Our numerical simulations and the econometric research we reviewed also make us conclude that workers' compensation insurance encourages workplace safety. Because increased workplace injuries raise a firm's insurance premiums workers' compensation acts as a tax on injuries. As with any tax employers can limit payments by reducing the activity taxed. If the cost of purchasing WC falls more than the cost of new safety equipment or safety education programs then the firm will pay more attention to workplace safety. When strongly experience rated workers' compensation insurance leads firms to the cost-benefit calculations necessary for cost effective safety expenditures.

One policy that would increase workplace safety is strengthening the dimensions of WC that mimic an injury rate tax. Experience rating of insurance premiums could be applied to both large and small firms and could to a greater degree reflect both frequency and severity of injuries. We would also allow employees to sue employers for negligence. Although they receive WC benefits for income lost and medical and rehabilitation expenses workers now bear the costs of the pain and suffering from industrial injuries. In a labor market with scarce information concerning job related health hazards and worker mobility costs wages might not create compensating differentials sufficient to give workers implicit insurance and firms sufficient incentives to have optimal safety programs. Allowing workers to sue employers for pain and suffering should improve the safety incentives under workers' compensation by making employers more economically aware of the total costs of workplace injuries.

NOTES

[1]Our discussion of the history of OSHA draws heavily on Viscusi (1986a; 1992, Chapter 14).

[2]An in-depth economic examination of the cotton dust standard is Viscusi (1992, Chapter 10).

[3]In 1987 there were about 20.9 million manufacturing workers with an average annual salary of $26,356. Workers' compensation insurance premiums were about 1.4 percent of payroll. The fatality rate in manufacturing during 1987 was 5/100,000, and the lost workday accident rate was 5.3/100. An upper end estimate of the value of a worker's life saved in manufacturing was $6.2 million and a lower end estimate of the value of a worker's life saved was $1 million (Moore and Viscusi 1990, p. 14; Viscusi 1993). An upper end estimate of the value of preventing a nonfatal manufacturing injury was $41,100 and a lower end estimate of the value of eliminating a nonfatal injury was $16,400 in 1987 dollars (Viscusi 1983, p. 99). We used the midpoint estimate of a 4.5 percent reduction in all injuries due to OSHA during the 1980s, the midpoint estimate of the benefit of a life saved of $3.6 million, and the midpoint estimate of the benefit of eliminating a nonfatal injury of $28,750. The estimated benefits from OSHA safety regulations are then (1) reduced workers' compensation insurance premiums ($26,356 × 20,935,000 × 0.014 × 0.045) = $348 million, (2) the value of reducing nonfatal injuries (20,935,000 × $28,750 × 0.052 × 0.045) = $1408 million, and (3) the value of lives saved (20,935,000 × $3,600,000 × 0.00005 × 0.045) = $170 million. Using a typical real interest rate for the United States of three percent yields an estimated total capitalized value for the benefits of OSHA of ($348 + $1408 + $170)/0.03 = $64 billion, which is about one-fifth the midpoint of the $100 to $500 billion estimated compliance costs.

[4]As mentioned the disability insurance provided workers under WC is largely confined to injuries due to accidents. Industrial diseases are covered under Social Security Disability Insurance, which gives employers no financial incentives to improve workplace health and safety.

[5]The reporting effect is one reason that claims for injuries that are tricky to verify medically, such as sprains and lower back pains, are higher on Monday mornings than on other days of the week (Smith 1990).

[6]Recent estimates are that a ten percent rise in a state's maximum workers' compensation benefits, an increase in weekly income support to injured workers of $24, would lower the average annual wage by about $83.

[7]According to our simulations of the previous chapter complete experience rating of all firms would lower injuries by about one percent.

REFERENCES

Associated Press. (1991a). "Fatal Plant Fire Leaves Children Without Parents," *Chicago Tribune* September 8: 18.

_____ (1991b). "OSHA Steps in to Help N.C. Enforce Safety Rules," *Boston Globe* October 24: 40.

_____ (1994). "Indiana Supreme Court Rules to Allow Injured Workers to Sue for Civil Damages," *Indiana Daily Student* June 30: 5.

Ball, K. (1992). "California First, Arkansas Last in Job Safety: U.S. Faulted," *Boston Globe* October 24: 40.

Bartel, A.P. and Thomas, L.G. (1985). "Direct and Indirect Effects of Regulation: A New Look at OSHA's Impact," *Journal of Law and Economics* 28(1): 1–26.

Bennett, J.A. (1991). "On-the-Job Poisoning," *Newsday* April 24: 84.

Berkowitz, E.D. and Berkowitz, M. (1985). "Challenges to Workers' Compensation: An Historical Analysis," In J.D. Worrall, eds., *Workers' Compensation Benefits: Adequacy, Equity, and Efficiency*, Ithaca, NY: Industrial Relations Press, 158–79.

Berkowitz, M. and Burton, J.F., Jr. (1970). "The Income-Maintenance Objective in Workmen's Compensation," *Industrial and Labor Relations Review* 24(1): 14–31.

Bernstein, H. (1991). "Fatal Fire May Stir OSHA Action," *Los Angeles Times* September 17: 3.

Berreth, C.A. (1992). "Workers' Compensation: State Enactments in 1991," *Monthly Labor Review* 115(1): 56–63.

_____ (1994). "Workers' Compensation Laws: Significant Changes in 1993," *Monthly Labor Review* 117(1): 53–64.

Brown, J.N. (1983). "Structural Estimation in Implicit Markets," In J. Triplett, ed., *Measurement of Labor Cost*, Chicago: University of Chicago Press for the National Bureau of Economic Research, 123–52.

Brown, J.N. and Rosen, H.S. (1982). "On the Estimation of Structural Hedonic Price Models," *Econometrica* 50(3): 765–68.

Brown, R.A. (1993). "Workers' Compensation: State Enactments in 1992," *Monthly Labor Review* 116(1): 50–56.

Burroughs, B. and Lubove, S.H. (1986). "Credibility Gap, Some Concerns Fudge Their Safety Records to Cut Insurance Costs," *Wall Street Journal*, December 2.

Butler, R.J. (1983). "Wage and Injury Rate Response to Shifting Levels of Workers' Compensation." In J. D. Worrall, ed., *Safety and the Workforce*, Ithaca, NY: Cornell University, ILR Press, 61–86.

Butler, R.J. and Worrall, J.D. (1983). "Workers' Compensation: Benefit and Injury Claims Rates in the Seventies," *Review of Economics and Statistics* 65(4): 580–89.

______ (1985). "Work Injury Compensation and the Duration of Nonwork Spells," *Economic Journal* 95(3): 714–24.

______ (1988). "Labor Market Theory and the Distribution of Workers' Compensation Losses," In P.S. Borba and D. Appel, eds., *Workers' Compensation Insurance Pricing — Current Programs and Proposed Reforms*, Boston: Kluwer Academic Publishers, 19–34.

______ (1991). "Claims Reporting and Risk Bearing Moral Hazard in Workers' Compensation," *Journal of Risk and Insurance*, 53(2): 191–204.

______ (1994)."Safety Through Experience Rating: A Review of the Evidence and Some New Findings," unpublished manuscript, Industrial Relations Center, University of Minnesota.

Chelius, J.R. (1974). "The Control of Industrial Accidents: Economic Theory and Empirical Evidence," *Law and Contemporary Problems* 38(4): 700–29.

_____ (1976). "Liability for Industrial Accidents: A Comparison of Negligence and Strict Liability Systems," *Journal of Legal Studies* 5(2): 293–309.

_____ (1977). *Workplace Safety and Health — The Role of Workers' Compensation*, Washington, DC: American Enterprise Institute for Public Policy Research.

_____ (1982). "The Influence of Workers' Compensation on Safety Incentives," *Industrial and Labor Relations Review* 35(2): 235–42.

_____ (1983). "The Incentive to Prevent Injuries," In J.D. Worrall, ed., *Safety and the Workforce*, Ithaca, NY: ILR Press, 154–60.

Chelius, J.R. and Kavanaugh, K. (1988). "Workers' Compensation and the Level of Occupational Injuries," *Journal of Risk and Insurance* 55(2): 315–23.

Chelius, J.R. and Smith, R.S. (1983). "Experience-Rating and Injury Prevention," In J.D. Worrall, ed., *Safety and the Workforce*, Ithaca, NY: Cornell University, ILR Press, 128–37.

_____ (1987). "Firm Size and Regulatory Compliance Costs: The Case of Workers' Compensation," *Journal of Policy Analysis and Management* 6(2): 193–206.

Cooke, W.N. and Gautschi, F.H., III (1981). "OSHA, Plant Safety Programs, and Injury Reduction," *Industrial Relations* 20(3): 245–257.

Countryman, G.L. (1994). "For Workers' Comp, Say No to Reform," *The New York Times* July 31: F9.

Curington, W.P. (1986). "Safety Regulation and Workplace Injuries," *Southern Economic Journal* 53(1): 51–72.

_____ (1994)."Compensation for Permanent Impairment and the Duration of Work Absence," *Journal of Human Resources* 29(3): 888-910.

Daily Labor Report (1992a). Washington, DC: Bureau of National Affairs, May 29.

_____ (1992b). Washington, DC: Bureau of National Affairs, July 8.

_____ (1994a). Washington, DC: Bureau of National Affairs, March 30.

_____ (1994b). Washington, DC: Bureau of National Affairs, April 28.

Darling-Hammond, L. and Kniesner, T.J. (1980). *The Law and Economics of Workers' Compensation*, Santa Monica, CA: The Rand Corporation.

Ehrlich, I. and Becker, G.S. (1972). "Market Insurance, Self-Insurance, and Self-Protection," *Journal of Political Economy* 80(4): 623–48.

Epple, D. (1987). "Hedonic Prices and Implicit Markets: Estimating Supply and Demand Functions for Differentiated Products," *Journal of Political Economy* 95(1): 59–80.

Executive Office of the President (1987). *Economic Report of the President*. Washington, DC: U.S. Government Printing Office.

Feldstein, M.S. (1977). "Social Insurance," *Public Policy* 25(1): 81–115.

Fishback, P.V. (1987). "Liability Rules and Accident Prevention in the Workplace: Empirical Evidence from the Early Twentieth Century," *Journal of Legal Studies* 16(2): 305–28.

Fleisher, B.M. and Kniesner, T.J. (1984). *Labor Economics — Theory, Evidence, and Policy*, Englewood Cliffs, NJ: Prentice-Hall, Inc.

Gray, W.B. and Scholz, J.T. (1993). "Does Regulatory Enforcement Work? A Panel Analysis of OSHA Enforcement," *Law and Society Review* 27(1): 177–213.

Greene, W.H. (1993). *Econometric Analysis, Second Edition*, New York: Macmillan Publishing Company.

Griliches, Z., ed. (1971). *Price Indexes and Quality Change*, Cambridge, MA: Harvard University Press.

Halpern, J. and Hausman, J. (1986). "Choice Under Uncertainty: Labor Supply and the Decision to Apply for Disability Insurance," In R. Blundell and I. Walker, eds., *Unemployment and Labor Supply*, Cambridge, UK: Cambridge University Press 294–302.

Hamermesh, D.S. (1993). *Labor Demand*, Princeton, NJ: Princeton University Press.

Harberger, A.C. (1962). "The Incidence of the Corporation Income Tax," *Journal of Political Economy* 70(3): 215–40.

Heckman, J.J., and Sedlacek, G.L. (1990). "Self-Selection and the Distribution of Hourly Wages," *Journal of Labor Economics* 8(1, part 2): S329–S363.

Hirschleifer, J., and Riley, J.C. (1992). *The Analytics of Uncertainty and Information*, Cambridge, UK: Cambridge University Press.

Hwang, H., Reed, W.R., and Hubbard, C. (1992). "Compensating Wage Differentials and Unobserved Productivity," *Journal of Political Economy* 100(41): 835–58.

Johnson, W.G. and Ondrich, J. (1990). "The Duration of Post-Injury Absences from Work," *Review of Economics and Statistics* 72(4): 578–86.

Johnson, W.G. and Wai, H.S. (1980). "Entitlement to Disability Benefits from More than One Public Program." Syracuse University Health Studies Program, Discussion Paper, March.

Kahn, S. and Lang, K. (1988). "Efficient Estimation of Structural Hedonic Systems," *International Economic Review* 29(1): 157–66.

Kasper, D.M. (1975). "An Alternative to Workmen's Compensation," *Industrial and Labor Relations Review* 28(4): 535–48.

Kniesner, T.J. and Leeth, J.D. (1988). "Simulating Hedonic Labor Market Models: Computational Issues and Policy Applications," *International Economic Review* 29(4): 755–90.

______ (1989a). "Can We Make OSHA and Workers' Compensation Insurance Interact More Effectively in Promoting Workplace Safety?" *Research in Labor Economics* 10: 1–51.

______ (1989b). "Separating the Reporting Effects from the Injury Rate Effects of Workers' Compensation Insurance: A Hedonic Simulation," *Industrial and Labor Relations Review* 42(2): 280–93.

______ (1991a). Improving Workplace Safety, Standards or Insurance?" *Regulation* Fall: 64–70.

______ (1991b). "Compensating Wage Differentials for Fatal Injury Risk in Australia, Japan, and the United States," *Journal of Risk and Uncertainty* 4(1): 75–90.

Krueger, A.B. (1990a). "Incentive Effects of Workers' Compensation Insurance," *Journal of Public Economics* 41(1): 73–99.

______ (1990b). "Workers' Compensation Insurance and the Duration of Workplace Injuries," National Bureau of Economic Research, Cambridge, MA, Working Paper No. 3253, February.

Krueger, A.B. and Burton, J.F., Jr. (1990). "The Employers' Costs of Workers' Compensation Insurance: Magnitudes, Determinants, and Public Policy," *Review of Economics and Statistics* 72(2): 228–40.

Leigh, J.P. (1985). "Analysis of Workers' Compensation Using Data on Individuals," *Industrial Relations* 24(2): 247–56.

Leigh, J.P. and Folsom, R.N. (1984). "Estimates of the Value of Accident Avoidance at the Job Depend on the Concavity of the Equalizing Differences Curve," *Quarterly Review of Economics and Business* 24(1): 56–66.

Lofgren, D.J. (1989). *Dangerous Premises, An Insider's View of OSHA Enforcement*, Ithaca, NY: ILR Press.

Mansur, A. and Whalley, J. (1984). "Numerical Specification of Applied General Equilibrium Analysis." In H.E. Scarf and J.B. Shoven, eds., *Applied General Equilibrium Analysis*, Cambridge: Cambridge University Press, 69–126.

May, L. (1991). "Plant Fire Held Unlikely to Spur Major Reforms," *Los Angeles Times* September 11: 4.

McCaffrey, D.P. (1983). "An Assessment of OSHA's Recent Effects on Injury Rates," *Journal of Human Resources* 18(1): 131–46.

McGarity, T.O. and Shapiro, S.A. (1993). *Workers at Risk: The Failed Promise of the Occupational Safety and Health Administration*, Westport, CN: Praeger Publishers.

Mendeloff, J. (1979). *Regulating Safety: An Economic and Political Analysis of Occupational Safety and Health Policy*, Cambridge, MA: MIT Press.

Moffitt, R. (1983). "An Economic Model of Welfare Stigma," *American Economic Review* 73(5): 1023–35.

Moore, M.J. and Viscusi, W.K. (1989). "Promoting Safety Through Workers' Compensation: The Efficacy and Net Wage Costs of Injury Insurance," *Rand Journal of Economics* 20(4): 499–515.

______ (1990). *Compensation Mechanisms for Job Risks — Wages, Workers' Compensation, and Product Liability*, Princeton: Princeton University Press.

National Safety Council. (1993). *Accident Facts, 1993 Edition*, Itasca, IL: National Safety Council.

Noble, K.B. (1986). "U.S. Cites Brooklyn Concern for Toxic Mercury Exposure," *New York Times* February 6: B2.

Nomani, A.Q. (1994). "Muffed Mission, Labor Secretary's Bid to Push Plant Safety Runs Into Skepticism," *Wall Street Journal*, August 19: A1.

Occupational Hazards. (1991). Cleveland, OH: Penton Publishing, December.

______ (1993a). Cleveland, OH: Penton Publishing, October.

______ (1993b). Cleveland, OH: Penton Publishing, December.

Oi, W. Y. (1973). "An Essay on Workmen's Compensation and Industrial Safety." In *Supplemental Studies for the National Commission on State Workmen's Compensation Laws, Vol. 1*. Washington, DC: U.S. Government Printing Office, 41–106.

Parsons, D.O. (1984). "Social Insurance with Imperfect State Verification: Income Insurance for the Disabled." The Ohio State University, Department of Economics, Working Paper No. 84-27, July.

Rea, S.A., Jr. (1981). "Workmen's Compensation and Occupational Safety Under Imperfect Information," *American Economic Review* 71(1): 80–93.

Regulation. (1991). "Making Sense of Safety," Washington, DC: Cato Institute, Fall.

Reynolds, L.G. (1974). *Labor Economics and Labor Relations*, 6th ed., Englewood Cliffs, NJ: Prentice-Hall.

Roel, R.E. (1988). "OSHA Cases: Prosecution Turning Into Frustration," *Newsday* July 29: 47.

Rosen, S. (1974). "Hedonic Prices and Implicit Markets: Product Differentiation in Pure Competition," *Journal of Political Economy* 82(1): 34–55.

______ (1986). "The Theory of Equalizing Differences." In O. C. Ashenfelter and R. Layard, eds., *The Handbook of Labor Economics*, Amsterdam: North Holland, 641–92.

Ruser, J.W. (1985). "Workers' Compensation Insurance, Experience-Rating, and Occupational Injuries," *Rand Journal of Economics* 16(4): 487–503.

______ (1991). "Workers' Compensation and Occupational Injuries and Illnesses," *Journal of Labor Economics* 9(4): 487–503.

______ (1993). "Workers' Compensation and the Distribution of Occupational Injuries," *Journal of Human Resources* 28(3): 593–617.

Ruser, J.W. and Smith, R.S. (1991). "Reestimating OSHA's Effects — Have the Data Changed?" *Journal of Human Resources* 26(2): 212–35.

Salwin, K.G. (1993). "Gore Task Force Considers Plan to Let Firms Do Their Own Safety Inspections," *Wall Street Journal*, September 2: A3.

Sattinger, M. (1977). "Compensating Wage Differences," *Journal of Economic Theory* 16(2): 496–503.

Schieber, S.J. (1982). *Social Security: Perspectives on Preserving the System*, Washington, DC: Employee Benefits Research Institute.

Scholz, J.T. and Gray, W.B. (1990). "OSHA Enforcement and Workplace Injuries: A Behavioral Approach to Risk Assessment," *Journal of Risk and Uncertainty* 19(3): 283–305.

Schwartz, D., Grundman, H., Kerns, W.L., Loeff, J., Bretz, J., Nelson, W., Bixby, A.K. and Simon, M.W. (1989). "Social Security Programs in the United States," *Social Security Bulletin* 52(7): 2–79.

Shoven, J.B. and Whalley, J. (1972). "A General Equilibrium Calculation of the Effects of Differential Taxation of Income from Capital in the U.S.," *Journal of Public Economics* 1(3): 281–321.

______ (1992). *Applying General Equilibrium*, New York: Cambridge University Press.

Smith, R.S. (1974). "The Feasibility of an 'Injury Tax' Approach to Occupational Safety," *Law and Contemporary Problems* 38(4): 730–44.

______ (1976). *The Occupational Safety and Health Act — Its Goals and Its Achievements*, Washington, DC: American Enterprise Institute for Public Policy Research.

______ (1979). "The Impact of OSHA Inspections on Manufacturing Injury Rates," *Journal of Human Resources* 14(2): 145–70.

______ (1990). "Mostly on Mondays: Is Workers' Compensation Covering Off-the-Job Injuries?" In P.S. Borba and D. Appel, eds., *Benefits, Costs, and Cycles in Workers' Compensation*, Boston: Kluwer Academic Publishers, 115–28.

______ (1992). "Have OSHA and Workers' Compensation Made the Workplace Safer?" In D. Lewin, O.S. Mitchell, and P.D. Sherer, eds., *Research Frontiers in Industrial Relations and Human Resources*, Madison, WI: Industrial Relations Research Association, University of Wisconsin, 557–86.

Smothers, R. (1991). "Twenty-Five Die, Many Reported Trapped, As Blaze Engulfs Carolina Plant," *New York Times* September 4: A1.

Taylor, P. (1991). "City Officials Urge Criminal Probe of Safety Lapses at Fatal Fire Scene," *Washington Post* September 7: A3.

Thaler, R. and Rosen, S. (1975). "The Value of Saving a Life: Evidence from the Labor Market." In N. Tereckyj, ed., *Household Production and Consumption*, New York: Columbia University Press, National Bureau of Economic Research, 265–97.

Time. (1994). April 18: 58.

Tinbergen, J. (1956). "On the Theory of Income Distribution," *Weltwirschaftliches Archiv* 77(2): 155–73.

Tinsley, L.C. (1991). "State Workers' Compensation: Legislation Enacted in 1990," *Monthly Labor Review* 114(1): 57–62.

Toscano, G. and Windau, J. (1993). "Fatal Work Injuries: Results from the 1992 National Census," *Monthly Labor Review* 116(10): 39–48.

Tye, L. (1991). "Poultry Plant Blaze Lights Era of Neglect," *Boston Globe* September 8: 1.

U.S. Department of Labor (1973). *Report on Occupational Safety and Health*, Washington, DC: U.S. Government Printing Office.

______ (1993). "News Release: Workplace Injuries and Illnesses in 1992," Bureau of Labor Statistics, Washington, DC, December 15.

Viscusi, W.K. (1979). "The Impact of Occupational Safety and Health Regulation," *The Bell Journal of Economics* 10(1): 117–40.

______ (1983). *Risk by Choice — Regulating Heath and Safety in the Workplace*, Cambridge, MA: Harvard University Press.

______ (1986a). "The Impact of Occupational Safety and Health Regulation, 1973-1983," *Rand Journal of Economics* 17(4): 567–80.

______ (1986b). "The Structure and Enforcement of Job Safety Regulation," *Law and Contemporary Problems* 49(4): 601–24.

______ (1986c). "Reforming OSHA Regulation of Workplace Risks," In L.W. Weiss and M.W. Klaus, eds., *Regulatory Reform: What Actually Happened*, Boston: Little, Brown, and Co., 234–68.

______ (1988). "Pain and Suffering in Product Liability Cases: Systematic Compensation or Capricious Awards?" *International Review of Law and Economics* 8(2): 203–20.

______ (1992). *Fatal Tradeoffs — Public and Private Responsibilities for Risk*, New York: Oxford University Press.

______ (1993). "The Value of Risks to Life and Health," *Journal of Economic Literature* 31(4): 1912–46.

Viscusi, W.K. and Evans, W.N. (1990). "Utility Functions That Depend on Health Status: Estimates and Economic Implications," *American Economic Review* 80(3): 353–74.

Viscusi, W.K. and Moore, M.J. (1987). "Workers' Compensation: Wage Effects, Benefit Inadequacies, and the Value of Health Losses," *Review of Economics and Statistics* 69(2): 249–61.

Viscusi, W.K. and O'Connor, C. (1984). "Adaptive Responses to Chemical Labeling: Are Workers Bayesian Decision Makers?" *American Economic Review* 74(5): 942–56.

Wall Street Journal. (1991). "AFL-CIO Asks U.S. to Police Employers in North Carolina," September 12: A8.

Watson, M.W. (1993). "Measures of Fit for Calibrated Models," *Journal of Political Economy* 101(6): 1011–41

Worrall, J.D. and Appel, D. (1982). "The Wage Replacement Rate and Benefit Utilization in Workers' Compensation Insurance," *Journal of Risk and Insurance* 49(3): 361–71.

Worrall, J.D. and Butler, R.J. (1985). "Benefits and Claim Duration," In J.D. Worrall and D. Appel, eds., *Workers' Compensation Benefits: Adequacy, Equity, and Efficiency*, Ithaca, NY: ILR Press, 57–70.

______ (1988). "Experience Rating Matters," In P.S. Borba and D. Appel, eds., *Workers' Compensation Insurance Pricing: Current Programs and Proposed Reforms*, Boston: Kluwer Academic Publishers, 81–94.

INDEX